The Hilbert Function of a Level Algebra

Memoirs
of the
American Mathematical Society

Number 872

The Hilbert Function of a Level Algebra

Anthony V. Geramita
Tadahito Harima
Juan C. Migliore
Yong Su Shin

March 2007 • Volume 186 • Number 872 (second of five numbers) • ISSN 0065-9266

American Mathematical Society
Providence, Rhode Island

2000 *Mathematics Subject Classification.* Primary 13D40, 13D02; Secondary 13C13, 13C40, 14C20.

Library of Congress Cataloging-in-Publication Data

The Hilbert function of a level algebra / Anthony V. Geramita . . . [et al.].
    p. cm. — (Memoirs of the American Mathematical Society, ISSN 0065-9266 ; no. 872)
    "Volume 186, number 872 (second of 5 numbers)."
    Includes bibliographical references.
    ISBN-13: 978-0-8218-3940-9 (alk. paper)
    ISBN-10: 0-8218-3940-3 (alk. paper)
    1. Algebra, Homological. 2. Characteristic functions. I. Geramita, A. V.
QA169.H535 2007
512'.64—dc22   2006047920

Memoirs of the American Mathematical Society

This journal is devoted entirely to research in pure and applied mathematics.

Subscription information. The 2007 subscription begins with volume 185 and consists of six mailings, each containing one or more numbers. Subscription prices for 2007 are US$649 list, US$519 institutional member. A late charge of 10% of the subscription price will be imposed on orders received from nonmembers after January 1 of the subscription year. Subscribers outside the United States and India must pay a postage surcharge of US$38; subscribers in India must pay a postage surcharge of US$43. Expedited delivery to destinations in North America US$53; elsewhere US$130. Each number may be ordered separately; *please specify number* when ordering an individual number. For prices and titles of recently released numbers, see the New Publications sections of the *Notices of the American Mathematical Society*.

Back number information. For back issues see the *AMS Catalog of Publications*.

Subscriptions and orders should be addressed to the American Mathematical Society, P. O. Box 845904, Boston, MA 02284-5904, USA. *All orders must be accompanied by payment.* Other correspondence should be addressed to 201 Charles Street, Providence, RI 02904-2294, USA.

Copying and reprinting. Individual readers of this publication, and nonprofit libraries acting for them, are permitted to make fair use of the material, such as to copy a chapter for use in teaching or research. Permission is granted to quote brief passages from this publication in reviews, provided the customary acknowledgment of the source is given.

Republication, systematic copying, or multiple reproduction of any material in this publication is permitted only under license from the American Mathematical Society. Requests for such permission should be addressed to the Acquisitions Department, American Mathematical Society, 201 Charles Street, Providence, Rhode Island 02904-2294, USA. Requests can also be made by e-mail to `reprint-permission@ams.org`.

Memoirs of the American Mathematical Society is published bimonthly (each volume consisting usually of more than one number) by the American Mathematical Society at 201 Charles Street, Providence, RI 02904-2294, USA. Periodicals postage paid at Providence, RI. Postmaster: Send address changes to Memoirs, American Mathematical Society, 201 Charles Street, Providence, RI 02904-2294, USA.

© 2007 by the American Mathematical Society. All rights reserved.
Copyright of this publication reverts to the public domain 28 years
after publication. Contact the AMS for copyright status.
This publication is indexed in *Science Citation Index*®, *SciSearch*®, *Research Alert*®,
CompuMath Citation Index®, *Current Contents*®/*Physical, Chemical & Earth Sciences*.
Printed in the United States of America.

∞ The paper used in this book is acid-free and falls within the guidelines
established to ensure permanence and durability.
Visit the AMS home page at `http://www.ams.org/`

10 9 8 7 6 5 4 3 2 1   12 11 10 09 08 07

Contents

Part 1. Nonexistence and Existence   1

Chapter 1. Introduction   3

Chapter 2. Numerical Conditions   7

Chapter 3. Homological Methods   13

Chapter 4. Some Refinements   27

Chapter 5. Constructing Artinian Level Algebras   31
    5.1. Inverse Systems   31
    5.2. Level Quotients of the Co-ordinate Rings of Points   37
    5.3. Constructing Level Algebras with the WLP   40
    5.4. The Linked-Sum Method   48

Chapter 6. Constructing Level Sets of Points   55
    6.1. Method 1: Intersection of Suitable Curves   55
    6.2. Method 2: Liaison Tricks   58
    6.3. Method 3: Union of suitable sets of points   62
    6.4. Method 4: General Sets of Points on Suitable curves   64

Chapter 7. Expected Behavior   67

Part 2. Appendix: A Classification of Codimension Three Level Algebras of Low Socle Degree   73

Appendix A. Introduction and Notation   75

Appendix B. Socle Degree 6 and Type 2   77

Appendix C. Socle Degree 5   93

Appendix D. Socle Degree 4   119

Appendix E. Socle Degree 3   129

Appendix F. Summary   133

Appendix. Bibliography   137

Abstract

Let R be a polynomial ring over an algebraically closed field and let A be a standard graded Cohen-Macaulay quotient of R. We say that A is a *level algebra* if the last module in the minimal free resolution of A (as R-module) is of the form $R(-s)^a$, where s and a are positive integers. When $a = 1$ these are also known as *Gorenstein algebras*.

The basic question addressed in this paper is: What can be the Hilbert Function of a level algebra? Our approach is to consider the question in several particular cases. E.g. when A is an Artinian algebra, or when A is the homogeneous coordinate ring of a reduced set of points, or when A satisfies the Weak Lefschetz Property.

We give new methods for showing that certain functions are NOT possible as the Hilbert function of a level algebra and we also give new methods to construct level algebras.

In a (rather long) appendix, we apply our results to give complete lists of all possible Hilbert functions in the case that the codimension of $A = 3$, s is small and a takes on certain fixed values.

Received by the editor January 26, 2005.

1991 *Mathematics Subject Classification*. Primary:13D40, 13D02; Secondary:13C13, 13C40, 14C20.

Key words and phrases. Level algebras, Hilbert function, liaison, link-sum, Weak Lefschetz property, Betti diagram, minimal free resolution.

*Supported in part by a grant from the Natural Sciences and Engineering Research Council (NSERC) of Canada and by the Ministero dell'Istruzione, dell'Università e della Ricerca (MURST) of Italy.

**This work was supported by two grants from Shikoku University and Hokkaido University of Education.

†Part of this work was done while this author was sponsored by the National Security Agency under Grant Number MDA904-03-1-0071.

‡This work was supported by a grant from Sungshin Women's University in 2002.

Part 1

Nonexistence and Existence

CHAPTER 1

Introduction

Let $R = k[x_1, \ldots, x_n] = \oplus_{i \geq 0} R_i$, k an algebraically closed field of characteristic 0, and let I be a homogeneous ideal of R, $A = R/I$. The *Hilbert function of A*, $\mathbf{H}_A : \mathbb{N} \to \mathbb{N}$, (or sometimes $\mathbf{H}(A, -)$) defined by:

$$\mathbf{H}_A(t) = \dim_k R_t - \dim_k I_t$$

has been much studied. In case I is the ideal of a subscheme, $\mathbb{X}$ of $\mathbb{P}^{n-1}$, (in which case the Hilbert function of $A = R/I$ is sometimes denoted $\mathbf{H}_\mathbb{X}(-)$ or $\mathbf{H}(\mathbb{X}, -)$) then this function contains a great deal of information about the geometry of this subscheme.

What possible functions arise in this context? This question was successfully considered by Macaulay in [**51**].

That solution was not, however, the end of the story. Many other, related, questions have also been considered:

i) What can $\mathbf{H}_A$ be if A is a domain? (see [**65**]);

ii) What can $\mathbf{H}_A$ be if $I = I_\mathbb{X}$ is the ideal of a reduced set of points, $\mathbb{X}$, in $\mathbb{P}^{n-1}$? (see [**25**]);

iii) What can $\mathbf{H}_A$ be if $I = I_\mathbb{X}$ is the ideal of a set of points, $\mathbb{X}$, which is the generic hyperplane section of a curve in $\mathbb{P}^n$? (see [**26**], [**34**], [**52**]);

iv) What can $\mathbf{H}_A$ be if A is a Gorenstein ring? (see [**3**], [**11**], [**16**], [**17**], [**28**], [**30**], [**32**], [**33**], [**37**], [**46**], [**55**], [**64**], [**71**]).

We can rephrase *iv)* above as follows: let R be as above and let I be a homogeneous ideal for which $\sqrt{I} = (x_1, \ldots, x_n)$. If $s+1$ is the least integer such that $(x_1, \ldots, x_n)^{s+1} \subseteq I$ then

$$A = k \oplus A_1 \oplus \cdots \oplus A_s \quad \text{where} \quad A_s \neq 0.$$

The *socle of A*, denoted $\text{soc}(A)$, is defined by

$$\text{soc}(A) := \text{ann}_A(m) \quad \text{where} \quad m = \oplus_{i=1}^s A_i.$$

Since m is a homogeneous ideal of A, $\text{soc}(A)$ is also a homogeneous ideal of A. Clearly, $A_s \subset \text{soc}(A)$.

Write

$$\text{soc}(A) = \mathfrak{A}_1 \oplus \cdots \oplus \mathfrak{A}_s \quad (\text{noting that } \mathfrak{A}_s = A_s)$$

and let $a_i = \dim_k(\mathfrak{A}_i)$. The integer vector

$$\mathbf{s} = s(A) = (a_1, \ldots, a_s)$$

is called the *socle vector* of A. Notice that $a_s = \dim_k \mathfrak{A}_s \neq 0$. We also call s the *socle degree* of A.

It is well-known that A is a Gorenstein ring if and only if $s(A) = (0, \ldots, 0, 1)$.

Another integer vector that we can associate to A is its *h-vector*, h, defined by
$$h = h(A) = (1, h_1, \ldots, h_s) = (1, \dim_k A_1, \ldots, \dim_k A_s)$$
which encodes the Hilbert function of A as a vector.

So, question iv) above becomes:

if $s(A) = (0, \ldots, 0, 1)$ what are the possibilities for $h(A)$?

We have a complete answer to question iv) in only two cases: when $h_1 = 2$ (well known) and when $h_1 = 3$ (see [**64**]).

In this monograph we will consider the following extension of question iv) to question $iv)'$:

$iv)'$ Let A be an algebra with socle degree s. If $s(A) = (0, \ldots, 0, c)$, with $c \geq 1$, what are the possibilities for $h(A) = (1, h_1, \ldots, h_s)$?

Algebras A for which $s(A) = (0, \ldots, 0, c)$, $c \geq 1$, are referred to in the literature as *level algebras of type c*, and their study was initiated by Stanley in [**63**]. Question $iv)'$ has also been considered in [**14**], [**20**], [**23**], [**43**], [**44**]. In particular, Iarrobino solved $iv)'$ in [**44**] for $h_1 = 2$ (see [**23**] and [**13**] for further references). Thus, our interest in $iv)'$ is in the case where $h_1 \geq 3$.

Level algebras have been studied in several different contexts. E.g., there is a strong connection between level algebras and *pure* simplicial complexes. More precisely, if Δ is a simplicial complex with n-vertices $(x_1, \ldots, x_n)$, let $k[\Delta]$ denote the Stanley-Reisner ring associated to Δ. Set $A_\Delta = k[\Delta]/(x_1^2, \ldots, x_n^2)$. Then the algebra A_Δ is level if and only if Δ is pure (see [**7**]).

Certain simplicial complexes also have level Stanley-Reisner rings. E.g., skeletons of Cohen-Macaulay complexes, triangulations of spheres and matroid complexes. Other examples come by considering the ideals of minors of a generic matrix. Also, for any $d \geq 1$, $n \geq 2$ and any t such that
$$\binom{d+n}{n} \leq t < \binom{d+1+n}{n} - \frac{1}{n}\binom{d+n}{n-1}$$
t general points in $\mathbb{P}^n$ have homogeneous coordinate ring which is level (see [**49**]).

This monograph is organized in the following way.

In Chapter 2 we make some preliminary definitions, recall some standard results about level algebras, and give our first results. Our first main result is to prove a decomposition for finite O-sequences which are the h-vectors of algebras with a given socle vector. This result (Theorem 2.10) extends and improves an analogous theorem of Stanley (Corollary 2.11). In Chapter 3 we reinterpret the notion of a level Artinian algebra homologically. Using this point of view we explain the combinatorial notion of Cancelation in Resolutions (first considered for level algebras in [**24**]). This simple idea becomes a powerful tool (thanks to a recent result of Peeva [**61**]) which we explore. In this chapter we also give some of our principal "non-level sequence" results. In Chapter 4 we use the homological point of view to define standard level algebras of any Krull dimension. We also recall the definition of the *Weak Lefschetz Property* (see [**71**]).

The weak and strong Lefschetz properties for Artinian algebras have an interesting history. Although R. Stanley has said (in a private communication) that he never explicitly mentions this property for arbitrary Gorenstein rings, he does assert that he was "morally aware of the concept since 1975" (see [**66**]). Several of his other papers ([**67**], [**68**]) amply support this view. We have always considered

Stanley as the "godfather" of this concept. The study of this property has also been taken up by several other authors for Gorenstein algebras (see [**35**], [**32**], [**55**], [**54**], [**71**]). Ours is the first systematic discussion of the WLP for level algebras (see Propositions 5.11, 5.15, 5.16, 5.18, 5.24, Corollary 5.17, and Example 6.18).

Chapters 5 and 6 are devoted to construction methods for level Artin algebras, reduced level algebras of positive Krull dimension and level algebras with the WLP. In section 5.1 we concentrate on the construction of Artin level algebras using Inverse Systems. In subsection 5.2 we explore level quotients of co-ordinate rings of sets of points in $\mathbb{P}^n$ and also explain some results of Boij [**8**] in this direction. In subsection 5.3 we concentrate on constructing level algebras which have the WLP. In subsection 5.4 we explain the "linked-sum" method for constructing level Artinian algebras. Since the "linked-sum" method requires us to have level algebras of positive Krull dimension readily at hand, we recall some results from [**23**] which explain how to construct (easily) useful sets of level points. Although the "linked-sum" method is very powerful for constructing level algebras we show (Remark 5.32) that it is not always possible to use it.

In Chapter 6 we consider the problem of constructing level sets of points in a more general way than we had considered earlier. We give four, essentially different, construction methods. Each of these methods is used to construct new examples of level algebras.

Chapter 7 is more speculative. There are natural candidates for level algebras, both at the Artinian level and at the points level, obtained by making "general" choices. We give a preliminary result and a conjecture, respectively, for these two situations.

In a (rather large) Appendix we give a complete list of the h-vectors of level Artin algebras of codimension 3 having socle degree ≤ 5 and of codimension 3, socle degree 6 and type 2. In all these cases we show that for each h-vector in our lists there is an example of a level algebra with that h-vector and having the WLP. For socle degree ≤ 4 and for type 2 in socle degrees 5 and 6 we show that every h-vector in our list is also the h-vector of a level set of points in $\mathbb{P}^3$.

We would like to take this opportunity to thank A. Iarrobino for his interest and support for this project and for generously sharing his insights about level algebras with us. We also would like to thank R. Stanley and B. Ulrich for an interesting discussion about our Theorem 2.10, and G. Dalzotto for the CoCoA program that was used to generate some of the initial lists in the appendix. It is also a pleasure to thank the MSRI for its kind hospitality to the first author during part of the writing of this monograph.

The authors dedicate their work on this book as follows: Geramita to his newly arrived (Dec. 2004) and much awaited first grandchild, Sophia Clara; Harima to his late father, Isamu Harima, who courageously endured his illness and died during the writing of this book; Migliore to his beloved parents, María Teresa Migliore and the late Francisco Migliore; and Shin to the memory of his late father, Sung-Ho Shin, and to his mother, Kyoung-Rye Kang, who has been fighting her illness in a hospital for a few years .

CHAPTER 2

Numerical Conditions

In this chapter we define a level sequence to be one that is the Hilbert function of some Artinian level algebra. We then give some elementary results and recall some facts from the literature. Our goal is to give several necessary numerical conditions for a sequence to be level. The main result of this chapter is an extension of Stanley's decomposition theorem (see Theorem 2.10).

Let $R = k[x_1, \ldots, x_n]$.

DEFINITION 2.1.
i) The sequence $\{h_i\}_{i \geq 0}$ (with $h_0 = 1$ and $h_1 \leq n$) is called an *O-sequence* if there is a homogeneous ideal $I \subset R$ such that if $A = R/I$ then $\mathbf{H}_A(i) = h_i$.
ii) In particular, the vector $h = (1, n, h_2, \ldots, h_s)$ is called an *O-sequence* if there is an Artinian quotient A of R whose h-vector is h.
iii) The vector $h = (1, n, h_2, \ldots, h_s)$ is called a *level sequence* if there is a level Artinian algebra having h-vector h. Moreover, we say that the sequence is a *Gorenstein sequence* if it is a level sequence with $h_s = 1$.

When $h_s \neq 0$ we say that $s + 1$ is the *length* of the sequence.

Some of the first results about level sequences were obtained by Stanley in [**63**]. He proved:

THEOREM 2.2. *Let $h = (1, h_1, \ldots, h_s)$ be a level sequence with $h_s \neq 0$. Then:*
  i) *for any t, $1 \leq t \leq s$, the sequence $h[t] := (1, h_1, \ldots, h_t)$ is also a level sequence;*
  ii) *for any i, j with $1 \leq i, j < s$ and with $i + j \leq s$ one has*
$$h_i \leq h_j h_{i+j};$$
  iii) *if $h_s = 1$ then $h_i = h_{s-i}$ for any i with $1 \leq i \leq s - 1$.*

PROOF. As for *i*), let $A = k \oplus A_1 \oplus \cdots \oplus A_s$ be a level algebra with h-vector h and consider
$$B = k \oplus A_1 \oplus \cdots \oplus A_t.$$
Clearly the h-vector of B is $h[t]$ and it is easy to see that B is also a level algebra.

As for *ii*), this is a simple consequence of the fact that the natural map $A_i \to \mathrm{Hom}_k(A_j, A_{i+j})$ is injective when A is level (see Lemma 2.7 for a generalization).

iii) is a consequence of the fact that, since A is Gorenstein, the multiplication map induces a bilinear form
$$A_i \times A_{s-i} \longrightarrow A_s \simeq k$$
which is a perfect pairing. The result follows. □

Stanley has one more very interesting observation to make about the h-vector of a level algebra. He shows that the h-vector of a level algebra has to "decompose" in a very particular way. After some preliminary definitions and lemmas, we give a proof of a generalization of Stanley's Decomposition Theorem which can be applied to Artinian algebras with *any* socle vector. As a corollary to that result we obtain a refinement of Stanley's original statement (undoubtedly known to Stanley, but not stated in his paper) that is very useful for our purposes. (See Remark 2.12 below.)

Let $A = \oplus_{i \geq 0} A_i = R/I$ be a standard graded k-algebra, where $R = k[x_1, \ldots, x_n]$. Let $0 \neq H \subsetneq A_s$ be a proper subspace of A_s. Following [44], we associate two ideals of A to the subspace H.

DEFINITION 2.3. Let $\mathfrak{A} = \mathfrak{A}(H)$ be defined by:

ai) if $d < s$ then $\mathfrak{A}_d = \{a \in A_d \mid aA_{s-d} \subseteq H\}$;
aii) if $d = s$ then $\mathfrak{A}_s = H$;
aiii) if $d > s$ then $\mathfrak{A}_d = HA_{d-s}$.

Then $\mathfrak{A}(H)$ is an ideal of A, called the *ancestor ideal* of H.
Let $\langle H : A \rangle$ be defined as:

bi) if $d < s$ then $\langle H : A \rangle_d = \{a \in A_d \mid aA_{s-d} \subseteq H\}$;
bii) if $d = s$ then $\langle H : A \rangle_s = H$;
biii) if $d > s$ then $\langle H : A \rangle_d = A_d$.

Then $\langle H : A \rangle$ is an ideal of A, called the *level ideal* of H.

REMARK 2.4. *i*) If A is Artinian of socle degree s and $0 \neq H \subsetneq A_s$ then $\mathfrak{A}(H) = \langle H : A \rangle$.

More generally,
ii) If $H \subseteq \mathrm{soc}(A)_s$ then $\mathfrak{A}(H)_t = 0$ for all $t > s$.

LEMMA 2.5. *Let $A = \oplus_{i \geq 0} A_i$ be a standard graded k-algebra and let $0 \neq H \subsetneq A_s$. Let $B = A/\langle H : A \rangle$. Then B is a level Artinian algebra with socle degree s and $B_s = A_s/H$.*

PROOF. The only thing that requires some comment is the statement that B is level.

So, suppose $\bar{a} \in B_t$, $t < s$, but $\bar{a}B_1 = 0$, i.e., $\bar{a} \in \mathrm{soc}(B)$. Then $aA_1 \subseteq \langle H : A \rangle_{t+1}$. Thus, $(aA_1)A_{s-(t+1)} \subseteq H$. But, since A is a standard algebra, $A_1 A_{s-(t+1)} = A_{s-t}$, so $aA_{s-t} \subseteq H$. I.e., $a \in \langle H : A \rangle_t$ and so $\bar{a} = 0$. □

LEMMA 2.6. *Let $A = \oplus_{i=1}^s A_i$ be a standard graded Artinian k-algebra for which $s(A) = (0, \ldots, 0, a_t, a_{t+1}, \ldots, a_s)$, where $a_s a_t \neq 0$. Let $0 \neq f \in \mathrm{soc}(A)_t$ and let $\mathfrak{A} = \mathfrak{A}(\langle f \rangle)$. Then $B = A/\mathfrak{A}$ is a standard graded Artinian k-algebra with*

$$s(B) = (0, \ldots, 0, a_t - 1, a_{t+1}, \ldots, a_s).$$

PROOF. It will be enough to prove that if $\bar{a} \in \mathrm{soc}(B)_u$ then $a \in \mathrm{soc}(A)_u$, since $f \in \mathrm{soc}(A)_t$ gives that $B_t = A_t/\langle f \rangle$ and $B_r = A_r$ for $r > t$.

Now, $\bar{a} \in \mathrm{soc}(B)_u \Rightarrow \bar{a}B_1 = 0$, i.e., $aA_1 \subseteq \mathfrak{A}_{u+1}$. Thus, if $u \geq t$, then $\mathfrak{A}_{u+1} = 0$ and so $a \in \mathrm{soc}(A)$. If $u < t$ then $aA_1 \subset \mathfrak{A}_{u+1} \Rightarrow (aA_1)(A_{t-(u+1)}) \subseteq \langle f \rangle$. Since A is a standard graded algebra, this implies that $aA_{t-u} \subset \langle f \rangle$, i.e., $a \in \mathfrak{A}_u$. Hence $\bar{a} = 0$ for $u < t$, and $s(B)$ is as claimed. □

LEMMA 2.7. *Let $A = k \oplus A_1 \oplus \cdots \oplus A_s = R/I$ be a level Artinian algebra with socle degree s and let J be any homogeneous ideal of A. Choose any indices $0 \leq i,j \leq s$ such that $i + j \leq s$ and consider the maps*

$$\varphi_{ij} : J_i \longrightarrow \mathrm{Hom}_k(A_j, J_{i+j})$$

defined as follows: if $x \in J_i$ then $\varphi_{ij}(x) \in \mathrm{Hom}(A_j, J_{i+j})$ where $[\varphi_{ij}(x)](a) = xa$. Then the maps φ_{ij} are all injective linear transformations.

PROOF. It is easy to see that the associations $x \to \varphi_{ij}(x)$ are linear transformations for all i,j, as in the statement of the Lemma.

We first consider the case $j = 0$.

Suppose that for some $i \leq s$ and some $x \neq 0$, $x \in J_i$, the map $\varphi_{i0}(x)$ is the zero linear transformation in $\mathrm{Hom}(k, J_i)$. But then we have $x = x \cdot 1 = 0$, which is nonsense.

Thus, the maps φ_{i0} are all injective.

Now, suppose that there are i, j such that $i + j \leq s$ and the map

$$\varphi_{ij} : J_i \longrightarrow \mathrm{Hom}(A_j, J_{i+j})$$

is not injective. Then, there is a smallest such j for which this is true (which must be ≥ 1) and which we will denote j'. Let $x \in J_i$ with $x \in \mathrm{Ker}\, \varphi_{ij'}$. Then

$$\begin{aligned}
\varphi_{ij'}(x) &\in \mathrm{Hom}(A_{j'}, J_{i+j'}) &&\text{is the zero map, i.e.,}\\
[\varphi_{ij'}(x)](a) &= 0 &&\text{for all } a \in A_{j'}, \text{ i.e.,}\\
xa &= 0 &&\text{for all } a \in A_{j'}.
\end{aligned}$$

Now $A_{j'} = A_1 A_{j'-1}$ so $xa = 0$ implies that $xua' = 0$ for all $u \in A_1$, $a' \in A_{j'-1}$. I.e., $xa' \in \mathrm{soc}(A)$.

But, A is level and $\deg(xa') = i + (j' - 1) < s$ and so $xa' = 0$ for any $a' \in A_{j'-1}$. But then $\varphi_{i,j'-1}(x)$ is the zero map, contradicting the minimality of j'. □

We now come to the key lemma in this circle of observations.

LEMMA 2.8. *Let A, f, $\mathfrak{A}$ be as in Lemma 2.6. Then the vector*

$$(\dim_k \mathfrak{A}_t = 1, \dim_k \mathfrak{A}_{t-1}, \ldots, \dim_k \mathfrak{A}_1, \dim_k \mathfrak{A}_0)$$

is an O-sequence which is the h-vector of a quotient of A.

PROOF. Let $C = k \oplus A_1 \oplus \cdots \oplus A_t$. Then clearly C is a level algebra and $f \in A_t$ is a socle element of C and $\mathfrak{A}$ is an ideal of C, since $\mathfrak{A}_r = 0$ for all $r > t$.

Let $E := \mathrm{Hom}_k(\mathfrak{A}, k)$. Then $E = E_0 \oplus \cdots \oplus E_t$ where $E_d := \mathrm{Hom}_k(\mathfrak{A}_{t-d}, k)$. Since $E_d = \mathfrak{A}_{t-d}^*$ we have that

$$\dim_k E_d = \dim_k \mathfrak{A}_{t-d}^* = \dim_k \mathfrak{A}_{t-d}.$$

In particular, $\dim_k E_0 = 1$.

Now, if $x \in C$ and $\varphi \in E$ and we define

$$(x \circ \varphi)(y) := \varphi(xy)$$

then E acquires the structure of a graded C-module. In fact,

Claim. If $0 \neq \xi \in E_0$ then $E \simeq C\xi$ as C-modules.

Notice that once the Claim is proved we will be done. To see why, consider the C-module homomorphism $\chi : C \to E$ defined by $\chi(a) = \xi a$. By the Claim, χ is surjective and so $E \simeq C/\mathrm{Ker}\chi$. By definition, the h-vector of $C/\mathrm{Ker}\chi$ is an O-sequence and that is precisely the sequence we are considering. The final statement of the Lemma is an immediate consequence of the fact that E is isomorphic to a quotient of C and C is a quotient of A.

So, it suffices to prove the Claim.

Proof of Claim. Clearly it suffices to show that $E_d = C_d \xi$ for every d, $0 \leq d \leq t$, i.e., that the maps $C_d \otimes E_0 \to E_d$, given by the C-module structure on E, are surjective; equivalently that the maps $E_d^* \to C_d^* \otimes E_0^*$ are injective.

Now,
$$E_d^* = \text{Hom}(\mathfrak{A}_{t-d}, k)^* = \text{Hom}(\text{Hom}(\mathfrak{A}_{t-d}, k), k) = \mathfrak{A}_{t-d}$$

and
$$\begin{aligned} C_d^* \otimes E_0^* &= \text{Hom}(C_d, k) \otimes \text{Hom}(\mathfrak{A}_t, k)^* \\ &= \text{Hom}(C_d, k) \otimes \mathfrak{A}_t \\ &= \text{Hom}(C_d, \mathfrak{A}_t). \end{aligned}$$

But, in Lemma 2.7 we showed that $\mathfrak{A}_{t-d} \to \text{Hom}(C_d, \mathfrak{A}_t)$ is injective for every d. That completes the proof of the Claim. $\square$

Before we state our extension of Stanley's Theorem, we need to make one more definition.

DEFINITION 2.9. Let $h = (1, h_1, \ldots, h_s)$ be an O-sequence. If $t > 0$ we define $h(t)$ to be the $t + (1 + s)$ vector,
$$h(t) = (0, \ldots, 0, 1, h_1, \ldots, h_s)$$
obtained from h by adjoining t zeroes to h at the beginning.

We say that $h(t)$ is the *t-shift* of h.

Now for the promised generalization of Stanley's Decomposition Theorem.

THEOREM 2.10. *Let A be an Artinian k-algebra with socle degree s for which*
$$h(A) = (1, h_1, \ldots, h_s) \text{ and } \text{soc}(A) = (0, \ldots, 0, a_t, \ldots, a_s)$$
where $a_t a_s \neq 0$. Then the reverse of $h(A)$ is the sum of:

 i) *a Gorenstein sequence $(1, b_1, \ldots, b_{s-1}, 1)$ (of length $s+1$) which is the h-vector of a quotient of A;*

 ii) *$h_s - 1$ O-sequences $(1, d_{j1}, \ldots, d_{jr_j})$, $j = 1, \ldots, h_s - 1$, each of which is the h-vector of a quotient of A;*

 iii) *a_j vectors $(j = t, \ldots, s-1)$ of the form $h_{j,r}(s-j)$, $r = 1, \ldots, a_j$, where $h_{j,r}(s-j)$ is the $(s-j)$-shift of an h-vector of a quotient of A.*

PROOF. We work by induction on $n = \sum_{i=t}^{s} a_i$ (the *Cohen-Macaulay (CM) type*) of A.

If $n = 1$ then A is Gorenstein and we are done.

So, suppose we are given A and the CM-type of A is n and we assume the theorem is true for all Artinian k-algebras of CM type $\leq n-1$. Choose $f \in \text{soc}(A)_t$, so f is a socle element of minimal degree in A, and let $\mathfrak{A} = \mathfrak{A}(\langle f \rangle)$. Consider the short exact sequence
$$0 \longrightarrow \mathfrak{A} \longrightarrow A \longrightarrow A/\mathfrak{A} \longrightarrow 0.$$
By Lemma 2.6, $B = A/\mathfrak{A}$ is an Artinian k-algebra with $\text{soc}(B) = (0, \ldots, a_t - 1, a_{t+1}, \ldots, a_s)$ and hence has CM type $= n - 1$. Thus, by induction, the theorem is true for B.

Comparing dimensions in every graded piece in the short exact sequence above we find that the reverse of the h-vector of A is the sum of the reverse of the h-vector of B and the vector

$$(0,\ldots,0,1=\dim_k \mathfrak{A}_t,\ldots,\dim_k \mathfrak{A}_1,0).$$

But, by Lemma 2.8, this last is the $(s-t)$ shift of an h-vector which is the h-vector of a quotient of A.

If we now apply the induction hypothesis to the ring B we are done. $\square$

As an immediate corollary we obtain,

COROLLARY 2.11. *Let* $h = (1, h_1, \ldots, h_s)$ *be a level sequence with* $h_s \geq 1$ *and let* A *be any level Artinian algebra with* $h(A) = h$. *Then the vector* $(h_s, h_{s-1}, \ldots, h_1, 1)$ *is the sum of:*
1) *a Gorenstein sequence* $(1, b_1, \ldots, b_{s-1}, 1)$ *(of length $s + 1$) which is the h-vector of a quotient of A and*
2) $h_s - 1$ *O-sequences* $(1, d_{j1}, \ldots, d_{jr_j})$, $j = 1, \ldots, h_s - 1$, *each of which is the h-vector of a quotient of A.*

REMARK 2.12. *a)* Stanley's original theorem in [**63**] simply states that the "reverse" of a level O-sequence (with first entry h_s) is the sum of h_s O-sequences. It does not mention that one of those sequences has to be a Gorenstein sequence nor does it mention that the Gorenstein sequence has to have length $s+1$. The fact that all the sequences have to be O-sequences of quotients of A is not mentioned either. The paper [**63**] does not offer any proof of the theorem, but it is clear that the line of argument we have given for Corollary 2.11 is probably exactly what Stanley had in mind. In fact, in private conversations with the first author, Stanley assured us that he was aware of these extra conditions.

In [**63**], Stanley points out that the reverse of $(1, 4, 2, 2)$ is not the sum of two O-sequences and so is not a level sequence. Note that in this case the reverse is not the sum of **ANY** two O-sequences, i.e., for this sequence one doesn't even need the extra conditions we mentioned above to eliminate it as a level sequence.

More telling, perhaps, is the sequence $h = (1, 3, 6, 5, 3, 2)$. The reverse of this sequence can be written as the sum of two O-sequences, namely $(2,3,5,6,3,1) = (1,3,5,6,3,1) + (1,0,0,0,0,0)$ even with one of them a Gorenstein O-sequence (the second). But, we cannot write this as a sum of two O-sequences which satisfy the conditions of Corollary 2.11. We'll see that below.

Notice also that $(2,5,10,6,3,1) = (1,1,1,1,1,1) + (1,4,9,5,2,0)$ gives the reverse of $h = (1, 3, 6, 10, 5, 2)$ as the sum of a Gorenstein sequence of the right length and another O-sequence. But, the second O-sequence in this decomposition is not the h-vector of a quotient of an algebra having h-vector h. In fact, as we'll see below, h is not a level sequence.

These two examples are a good illustration that the extra conditions mentioned in Corollary 2.11 are very useful.

b) A simple consequence of Theorem 2.10 is that if we let

$$\tilde{h} = (1, \tilde{h}_1, \ldots, \tilde{h}_r), \ \ r \leq s$$

be any one of the O-sequences in the decomposition promised by the theorem, then $\tilde{h}_1 \leq h_1$.

This fact can be used profitably to show that certain O-sequences are not level sequences.

E.g., consider any O-sequence $(1, 3, \ldots, 10, 5, 2)$. A simple check of the possibilities shows that the reverse of such a sequence cannot be written as the sum of two O-sequences, each with second entry ≤ 3, and satisfying the conditions 1) and 2) of Corollary 2.11.

Similar observations serve to show than any O-sequence of the following type:

$$(1, 3, \ldots, 2, 2), \quad (1, 3, \ldots, \geq 8, 4, 2), \quad (1, 3, \ldots, \geq 10, 5, 2),$$
$$(1, 3, 6, \ldots, 4, 3, 2), \quad (1, 3, 6, \ldots, 5, 3, 2), \quad (1, 3, 6, \ldots, 5, 5, 5, 2)$$

is not a level O-sequence.

c) Corollary 2.11 is also useful in other ways.

Consider the O-sequence $(1, 3, 6, 5, 4, 2)$ and suppose we want to investigate if it is a level sequence. There are only two ways that $(2, 4, 5, 6, 3, 1)$ can be written as the sum of an O-sequence and a Gorenstein sequence of length 6, namely

$$(1, 1, 1, 1, 1, 1) + (1, 3, 4, 5, 2, 0)$$

and

$$(1, 2, 2, 2, 2, 1) + (1, 2, 3, 4, 1, 0).$$

But notice that, as far as the first decomposition is concerned, $(1, 3, 4, 5, 2)$ cannot be the h-vector of a quotient of any level algebra A with h-vector $(1, 3, 6, 5, 4, 2)$. For, if it were, A would have an ideal I for which $I_2 \neq 0$ and $I_3 = 0$. I.e., $A_1 I_2 = 0$. I.e., $I_2 \subset \text{soc}(A)$, which is impossible.

So, we have only one way to satisfy the conditions of Corollary 2.11.

We'll see later that $(1, 3, 6, 5, 4, 2)$ is NOT a level O-sequence by showing that the unique possibility above cannot occur.

There are different ways we can use Corollary 2.11, in conjunction with other results, to show that certain O-sequences cannot be level sequences.

EXAMPLE 2.13. In this example we will show that there are no Artinian level algebras of socle degree s with h-vector $h = (1, 3, \alpha, \ldots, 5, 3, 2)$.

By Macaulay's theorem the only possibilities for α are 3, 4, 5 and 6. With a small amount of work one can show that the only decomposition that will work with Corollary 2.11 occurs when $\alpha = 5$ and

$$h = (1, 0, \ldots, 0) + (1, 3, 5, \ldots, 5, 3, 1)$$

where the second O-sequence must be a Gorenstein sequence of length $s + 1$.

Now suppose that $A = k[x_1, x_2, x_3]/I$ is an Artinian level algebra with h-vector h. Then the proof of Theorem 2.10 says that there is an ideal J where $I \subset J$ and $B = k[x_1, x_2, x_3]/J$ has h-vector the Gorenstein sequence above.

But by [[**29**], Theorem 2.17] (see also [**17**]) any Gorenstein ideal with that h-vector is generated in degrees $\leq s - 1$. But, J is generated by I and one more element of degree s. Since I and J agree in degrees $\leq s - 1$ we must conclude that $I = J$ and that is a contradiction.

CHAPTER 3

Homological Methods

There is another way to think about level Artinian algebras which is very important. Let $A = k \oplus A_1 \oplus \cdots \oplus A_s$ be any Artinian quotient of $R = k[x_1, \ldots, x_n]$ and let $\text{soc}(A) = \mathfrak{A}_1 \oplus \cdots \oplus \mathfrak{A}_s$ be as above. The only structure that the ideal $\text{soc}(A)$ has is that of a graded k-vector space. So, it is the socle vector of A, $s(A) = (a_1, \ldots, a_s)$ which actually records the dimensions of the graded pieces of this vector space.

We have that $A = R/I$ where I is a homogeneous ideal of R having height n and so A has a minimal graded free resolution $\mathbb{F}$, as R-module, of the form:

$$\mathbb{F}: \quad 0 \to \mathcal{F}_{n-1} \to \cdots \to \mathcal{F}_1 \to \mathcal{F}_0 \to R \to A \to 0$$

where the $\mathcal{F}_j$ are each free graded R-modules. In fact,

$$\mathcal{F}_j := \oplus_{t=1}^{r_j} R(-(j+1+t))^{\beta_{j,j+1+t}}, \ t \geq 0.$$

The numbers $\{\beta_{j,i}\}$, for fixed j, $0 \leq j \leq n-1$, are called the jth *graded Betti numbers* of the ideal I. It is well-known that the socle vector of A, $s(A) = (a_1, \ldots, a_s)$, is related to the $(n-1)^{st}$ graded Betti numbers as follows:

$$\beta_{n-1,n+i} = a_i.$$

It follows that A is a level algebra if and only if $\beta_{n-1,n+i} = 0$ for all $i \neq s$.

For I as above, the *Betti diagram* of R/I is a useful device to encode the graded Betti numbers of R/I (and hence of I). It is constructed as follows:

$$\begin{array}{c|ccccc} & 0 & 1 & \cdots & n-1 \\ \hline 0 & 1 & 0 & 0 & \cdots & 0 \\ 1 & 0 & * & * & \cdots & * \\ \vdots & \vdots & \vdots & \vdots & \vdots & \vdots \\ t & 0 & \beta_{0,t+1} & \beta_{1,t+2} & * & \beta_{n-1,t+n} \\ \vdots & \vdots & \vdots & \vdots & \vdots & \vdots \\ d-2 & 0 & \beta_{0,d-1} & \beta_{1,d} & * & \beta_{n-1,d-2+n} \\ d-1 & 0 & \beta_{0,d} & \beta_{1,d+1} & * & \beta_{n-1,d-1+n} \\ d & 0 & \beta_{0,d+1} & \beta_{1,d+2} & * & \beta_{n-1,d+n} \\ \vdots & \vdots & \vdots & \vdots & \vdots \end{array}$$

How can one use the relationship between the socle and the graded Betti numbers to discover if $h = (1, h_1, \ldots, h_s)$ is a level sequence? In the paper [24] the authors offered the following observations: it is a well-known theorem of Bigatti-Hulett-Pardue ([4], [41], [60]) that, given any O-sequence h, as above, there is an *extremal* algebra $\mathcal{E} = R/\mathcal{L}$ with $h(\mathcal{E}) = h$ and j^{th} graded Betti numbers $\{\varepsilon_{j,i}\}$ where $0 \leq j \leq n-1$ and $j+1 \leq i < j+1+r_j(\mathcal{E})$ having the following property:

if $B = R/I$ is *any* other graded k-algebra with $h(B) = h$ then the jth graded Betti numbers of I, call them $\{\tau_{j,i}\}$, must satisfy

$$\tau_{j,i} \leq \varepsilon_{j,i} \tag{3.1}$$

for all j and every i.

The ideal that Bigatti and Hulett use (in characteristic 0) is the lex segment ideal, $\mathcal{L}$ of R, for which $R/\mathcal{L}$ has h-vector h.

The simple observation of [**24**] is that, in order for h to support a level algebra, we have to be able to *cancel* all extraneous Betti numbers from the last free summand in the Bigatti-Hulett-Pardue (BHP) resolution. Moreover (following a recent observation of Peeva [**61**]) the only way we can "cancel" graded Betti numbers is if there are the same graded Betti numbers in the adjacent free modules of the BHP resolution (something obvious in case $n = 3$).

Let's illustrate this observation with a simple example.

EXAMPLE 3.1. Let $h = (1, 3, 4, 5, 4, 4, 2)$ be an O-sequence. Then, the extremal algebra $\mathcal{E} = R[x_1, x_2, x_3]/I$ with h-vector h above, has Betti diagram:

$$\begin{array}{c} \\ 0 \\ 1 \\ 2 \\ 3 \\ 4 \\ 5 \\ 6 \end{array} \begin{array}{c} \begin{array}{cccc} & 0 & 1 & 2 \end{array} \\ \begin{pmatrix} 1 & 0 & 0 & 0 \\ 0 & 2 & 1 & 0 \\ 0 & 0 & 0 & 0 \\ 0 & 2 & 3 & 1 \\ 0 & 0 & 0 & 0 \\ 0 & 2 & 4 & 2 \\ 0 & 2 & 4 & 2 \end{pmatrix} \end{array}$$

The only way that h could support a level algebra is if we could cancel all the non-zero terms in the last column except the final 2. The only way we could cancel something from the last column is if there is something in the next last column, and one row lower, which allows the cancelation. E.g., the 2 in the column labeled 2 of row 5 could (numerically) be canceled against the 4 in the column labeled 1 in row 6. However, the 1 in the column labeled 2 of row 3 cannot be canceled as the 4th row has a 0 in the column labeled 1. In other words, no matter what algebra we find with h-vector $(1, 3, 4, 5, 4, 4, 2)$, that algebra must have a one dimensional socle in degree 3.

Of course, even when cancelation is numerically possible, there is no guarantee that a level algebra will exist with a given h-vector. But still, the fact of non-cancelation (as we'll see later) permits us to eliminate many h-vectors as potential level sequences.

We were spurred on to thinking about this by a remark of Cho and Iarrobino in [**14**]. They indicated how h-vectors which contain certain subsequences could not be level sequences. They offered no explicit proof for their statement but indicated that their observation was a consequence of a theorem of Gotzmann.

In attempting to understand this remark we came up with a different explanation which uses a celebrated theorem of Eliahou-Kervaire. Since our approach is different and not only permits us to reprove the result of Cho and Iarrobino but also prove other results not mentioned by them, we think it worthwhile to indicate our reasoning.

Eliahou and Kervaire [**19**] studied minimal free resolutions of certain monomial ideals. We recall some of their notation and results now.

DEFINITION 3.2. Let $T \in R = k[x_1, \ldots, x_n]$ be a *term* of R. Then

$$m(T) := \max\{i \mid x_i \text{ divides } T\}.$$

i.e., $m(T)$ is the largest index of an indeterminate that divides T.

THEOREM 3.3 (Eliahou-Kervaire). *Let I be a stable monomial ideal of R (e.g. a lex segment ideal). Denote by $\mathcal{G}(I)$ the set of minimal (monomial) generators of I and by $\mathcal{G}(I)_d$ the elements of that set which have degree d. Then*

$$\beta_{q,i} = \sum_{T \in \mathcal{G}(I)_{i-q}} \binom{m(T) - 1}{q}$$

This wonderful theorem thus gives all the graded Betti numbers of a stable monomial ideal just from an intimate knowledge of the generators of that ideal. In the case of the lex segment ideal with a given Hilbert function H, we know - in theory - all the generators of the ideal and so can write down the graded Betti numbers algorithmically. Since the ideal that Bigatti and Hulett used to construct an extremal algebra with given h-vector is a lex segment ideal, we may apply this result to those ideals.

It is a simple consequence of the Eliahou-Kervaire theorem that if I is a stable monomial ideal which has NO generators of degree d then $\beta_{q,i} = 0$ whenever $i - q = d$.

Using the Betti diagram we can restate that observation: if I is a stable monomial ideal which has no generators of degree d (i.e., $\beta_{0,d} = 0$) then the entire row of the Betti diagram beginning with $\beta_{0,d}$ is zero.

So, a natural question is: when can we say, for a stable ideal I, that I has no generators of degree d? Very generally we know that I has no minimal generators of degree $d \Leftrightarrow R_1 I_{d-1} = I_d$.

A particular case in which we can be sure that $R_1 I_{d-1} = I_d$ is when the ideal I grows *minimally* in the passage from degree $d-1$ to degree d. I.e., if

$$h_{d-1} = \dim_k(R_{d-1}/I_{d-1}) \quad \text{and} \quad h_d = \dim_k(R_d/I_d)$$

then I has no minimal generators of degree d if (in the Macaulay notation - see [**51**])

(3.2) $$h_{d-1}^{<d-1>} = h_d$$

i.e., the passage from h_{d-1} to h_d is maximal growth (in the sense of Macaulay).

Thus, if we combine the result of Bigatti-Hulett and Pardue with that of Eliahou and Kervaire, we see that ANY algebra with Hilbert function satisfying (3.2) above must have the $(d-1)^{st}$ row of its Betti diagram identically 0.

This sets us up for the following result.

THEOREM 3.4 (Cho-Iarrobino). *Let h_{d-2}, h_{d-1}, h_d be three non-zero integers such that*

$$h_d = h_{d-1}^{<d-1>} \quad \text{and} \quad h_{d-1} = h_{d-2}^{<d-2>}.$$

Let I be any ideal in $R = k[x_1, \ldots, x_n]$ such that the Hilbert function of R/I satisfies
$$\begin{aligned} \mathbf{H}(R/I, d-2) &= h_{d-2} + \varepsilon, \quad \varepsilon \geq 0 \\ \mathbf{H}(R/I, d-1) &= h_{d-1}, \\ \mathbf{H}(R/I, d) &= h_d. \end{aligned}$$

Then, the ring R/I has socle of dimension ε in degree $d-2$. Consequently, if I has minimal free resolution $\mathbb{F}$, as above, then
$$\beta_{n-1, d+n-2} = \varepsilon.$$

PROOF. First note that it will be enough to prove the theorem for the Artinian ring
$$A = \frac{(R/I)}{(R/I)_{\geq d+1}}.$$
The reason is that both A and R/I agree in degree $\leq d$ and hence the socle of one in degree $d-2$ is the same as the socle of the other in that degree.

So, from now on we will assume that I is an Artinian homogeneous ideal.

Now, suppose further that I is a *lex-segment* ideal whose Hilbert function is as required in degrees $d-2$, $d-1$ and d.

If $\varepsilon = 0$ then, by the discussion above, I has no generators of degree $d-1$ and so $\beta_{0,d-1} = 0$. Consequently that entire row of the Betti diagram is $= 0$ and (in particular) $\beta_{n-1,d+n-2} = 0$.

Now suppose that $\varepsilon > 0$. By hypothesis, I has no generators in degree d, so if $\overline{m}$ is the largest monomial in I_{d-1} then $x_n \overline{m}$ is the last monomial in I_d. Let V be the lex segment in degree $d-2$ of codimension h_{d-2}. Then, by assumption, I_{d-2} is of codimension ε in V.

Since h_{d-2}, h_{d-1} is maximal growth we have that $R_1 V = I_{d-1}$. Now, the monomials of $I_{d-1} \backslash R_1 I_{d-2}$ are precisely the set of minimal generators of I in degree $d-1$. Since our interest is in $\beta_{n-1,d+n-2}$ we need to show that
$$\sum_{T \in \mathcal{G}(I)_{d-1}} \binom{m(T)-1}{n-1} = \varepsilon.$$

To do that we begin by showing that among the elements of $I_{d-1} \backslash R_1 I_{d-2}$ there are at least ε monomials T which are divisible by x_n, for in that case we will have, for those monomials, that $m(T) = n$ and thus each monomial will contribute 1 to the sum above.

But this is clear, since if $m_1, \ldots, m_\varepsilon$ are the distinct monomials of $V \backslash I_{d-2}$ then $x_n m_1, \ldots, x_n m_\varepsilon$ are in $I_{d-1} \backslash R_1 I_{d-2}$ and we are done.

To finish the argument for the lex segment ideal it suffices to show that (in the notation above) $x_n m_1, \ldots, x_n m_\varepsilon$ are the *only* monomials divisible by x_n in $I_{d-1} \backslash R_1 I_{d-2}$.

But, suppose that $m' \in R_{d-1} \backslash R_1 I_{d-2}$ and $x_n \mid m'$, i.e., $m' = x_n m''$. Then, since h_{d-2}, h_{d-1} is maximal growth, we must have $m'' \in V$. But, if $m'' \notin V \backslash I_{d-2}$ then $m'' \in I_{d-2}$ and hence $x_n m'' \in R_1 I_{d-2}$ which is a contradiction. So, $x_n m_1, \ldots, x_n m_\varepsilon$ are the only monomials among the generators of I in degree $d-1$ which are divisible by x_n and so $\beta_{n-1,d+n-2} = \varepsilon$ exactly.

Now, using the relationship between the $(n-1)^{st}$ graded Betti numbers of I and the socle of R/I, we conclude that R/I has socle precisely of dimension ε in degree $d-2$.

Thus, the $(d-2)^{nd}$ and $(d-1)^{st}$ rows of the Betti diagram of R/I look like:

$d-2:$ 0 $\beta_{0,d-1}$ $\cdots$ $\beta_{n-2,n+d-3}$ $\beta_{n-1,n+d-2} = \varepsilon,$

$d-1:$ 0 $\beta_{0,d} = 0$ $\cdots$ $\beta_{n-2,n+d-2} = 0$ $\beta_{n-1,n+d-1} = 0.$

Now, let I be *any* ideal whose Hilbert function agrees with that of the lex-segment ideal discussed above. By the Bigatti-Hulett-Pardue theorem, the Betti numbers in the resolution of R/I come from cancellation of Betti numbers in the resolution above. By the Peeva Cancelation Theorem, that cancelation can only occur in an adjacent free module in the resolution. Hence, in the particular case when we want to cancel something in the last term of the resolution it follows that the cancelation can only occur against something in the penultimate term. Thus, since $\beta_{n-2,n+d-2} = 0$ in the resolution of the lex segment ideal, that graded Betti number must be zero also in the resolution of our ideal. This implies that for our ideal, we must also have $\beta_{n-1,n+d-2} = \varepsilon$. Thus, any ideal I whose Hilbert function is as required, must have socle of dimension precisely ε in degree $d-2$. □

COROLLARY 3.5. *Let* $h = (1, h_1, \ldots, h_t)$ *be the h-vector of an Artinian quotient of* R *with socle degree* t. *Suppose that for some* $d-2 < t$ *we have three integers* h'_{d-2}, h'_{d-1} *and* h'_d *for which:*

$$h'_d = {h'_{d-1}}^{<d-1>} \quad \text{and} \quad h'_{d-1} = {h'_{d-2}}^{<d-2>}$$

and such that

$$h_{d-2} = h'_{d-2} + \varepsilon, \quad \varepsilon > 0, \quad h_{d-1} = h'_{d-1} \quad \text{and} \quad h_d = h'_d.$$

Then h *is not a level O-sequence.*

PROOF. From the theorem we have that any algebra with h-vector as given must have an ε-dimensional socle in degree $d-2$ and hence cannot be level. □

EXAMPLE 3.6. One can use this result to eliminate many possible O-sequences as level sequences. E.g., any h-vector $h = (1, h_1, \ldots, h_s)$ of an Artinian algebra of socle degree s for which $h_d = h_{d+1} = a < d$ and $h_{d-1} > a$ cannot be a level O-sequence. For example, anything of the form $(1, 3, \ldots, *_d, 2, 2, \ldots)$ is impossible as a level O-sequence whenever $*_d > 2$ and $d \geq 2$. In fact, we'll see that even the sequence $(1, 3, 2, \ldots, 2, 2)$ is impossible (see Proposition 3.8, b) below).

REMARK 3.7. We want to emphasize that Theorem 3.4 is a "non-cancelation" result of the type that we alluded to earlier. If $h = (1, h_1, \ldots, h_s)$ is the h-vector of an Artinian algebra of socle degree s satisfying the hypothesis of Theorem 3.4, then the proof of that theorem really shows that there are two rows in the BHP resolution associated to h which are of the form:

$d-2:$ 0 $\beta_{0,d-1}$ $\cdots$ $\beta_{n-2,n+d-3}$ $\beta_{n-1,n+d-2},$

$d-1:$ 0 $\beta_{0,d}$ $\cdots$ $\beta_{n-2,n+d-2}$ $\beta_{n-1,n+d-1},$

where

(3.3) $\qquad d-2 < s \quad \text{and} \quad \beta_{n-1,n+d-2} > \beta_{n-2,n+d-2}.$

(In the case of Theorem 3.4 we had $\beta_{n-1,n+d-2} > 0$ and $\beta_{n-2,n+d-2} = 0$.)

But, the important thing in Theorem 3.4 was really the information in (3.3).

Since, as we mentioned earlier, there is an algorithm to calculate *all* the graded Betti numbers in the BHP resolution starting only with h, it is possible to use (3.3) to easily eliminate many h-vectors as potential level sequences.

In fact, such an algorithm has been implemented in CoCoA [**12**] by E. Carlini and M. Stewart (Type Vector Package). We have used that programme extensively in the Appendix.

We now give some other results, similar to that of Theorem 3.4, which rely on Remark 3.7 and are not mentioned in [**14**].

PROPOSITION 3.8. *Let $h = (1, n, h_2, \ldots, h_s)$ be the h-vector of an Artinian algebra with socle degree s. Then h is **not** a level sequence in each of the following cases:*

  a) $h_d = h_{d+1} = p \leq d - 1$ and $h_{d-1} > p$;
  b) $h_d = h_{d+1} = p = d$ and $h_{d-1} > p = d$;
  c) $h_d = h_{d+1} = p = d + 1$ and $h_{d-1} > d + 1 = p$;
  d) $h_d = h_{d+1} = p \leq 2d$ and $h_{d-1} \geq p + n$ and $d \geq n + 2$.

PROOF. For a) note that p p p in degrees $d-1, d, d+1$ is maximal growth so this is an immediate consequence of Corollary 3.5 and generalizes Example 3.6.

We will need the following information both for the proof of b) and c), so we record it now.

The last d monomials of degree d in R are
$$x_{n-1}^{d-1}x_n, \ x_{n-1}^{d-2}x_n^2, \cdots, x_{n-1}x_n^{d-1}, \ x_n^d$$
and the last $d+2$ monomials in R_{d-1} are
$$x_{n-2}x_{n-1}x_n^{d-3}, x_{n-2}x_n^{d-2}, x_{n-1}^{d-1}, \ x_{n-1}^{d-2}x_n, \cdots, x_{n-1}x_n^{d-2}, \ x_n^{d-1}.$$

For b) note that the sequence d d in degrees $d, d+1$ is maximal growth. So, if I is the lex-segment ideal in R associated to h then I has no generators in degree $d+1$, i.e., $\beta_{0,d+1} = 0$.

We claim that there is a socle element in R/I of degree $d-1$, i.e $\beta_{n-1,n+d-1} > \beta_{n-2,n+d-1}$. But, since $\beta_{0,d+1} = 0$ this implies that $\beta_{n-2,n+d-1} = 0$ also. So, it's enough to show that $\beta_{n-1,n+d-1} \neq 0$.

Now,
$$\beta_{n-1,d+n-1} = \sum_{T \in \mathcal{G}(I)_d} \binom{m(T) - 1}{n - 1}$$

and so it will be enough to show that $m(T) = n$ for at least one generator of I of degree d.

Our assumption is that $h_d = d$, so the monomials not in I_d are the last d monomials of R_d, i.e., those listed above.

Our other assumption in this case is that $h_{d-1} > d$. So, we have (at least) that the last $d+1$ monomials of R_{d-1} (listed above) are not in I_{d-1}.

Now, consider $x_{n-2}x_n^{d-1} \in R_d$. This monomial is in I_d but it is not in R_1I_{d-1}. For if it were we would have either x_n^{d-1} or $x_{n-2}x_n^{d-2}$ in I_{d-1} and that is not the case.

Thus $x_{n-2}x_n^{d-1} \in \mathcal{G}(I)_d$ and $m(x_{n-2}x_n^{d-1}) = n$ and we are done in this case.

As for c), again we'll show that $\beta_{n-1,n+d-1} > \beta_{n-2,n+d-1}$ for the lex segment ideal associated to h.

Since $(d+1)^{<d>} = d+2$ we see that the lex segment ideal has exactly one generator in degree $d+1$ and that is x_{n-1}^{d+1}. For this generator $m(T) = n-1$ and so

$$\beta_{n-2,n+d-1} = \sum_{T \in \mathcal{G}(I)_{d+1}} \binom{m(T)-1}{n-2} = 1.$$

We turn now to

$$\beta_{n-1,n+d-1} = \sum_{T \in \mathcal{G}(I)_d} \binom{m(T)-1}{n-1}$$

and this requires us to investigate the generators of I of degree d. Since $h_{d-1} > d+1$, by hypothesis, this means that there are at least two minimal generators of I having degree d.

We claim that both $x_{n-2}x_{n-1}x_n^{d-2}$ and $x_{n-2}x_n^{d-1}$ are minimal generators of I. Since both of these involve x_n we will be done once this claim is proved.

Since both are in I_d, it suffices to show that neither is in $R_1 I_{d-1}$. The argument is as in b) and so we are done in this case as well.

We now consider part d) of this Proposition. We can assume, without loss of generality, that $p \geq d+2$ since the other cases have already been considered above.

Note that since $d+2 \leq p \leq 2d$ then $p = d+1+r$ with $r \leq d-1$. Thus the d-binomial expansion of p is:

$$p = \binom{d+1}{d} + \binom{d-1}{d-1} + \cdots + \binom{d-r}{d-r}$$

and hence $p^{<d>} = \binom{d+2}{d+1} + r = d+2+r = p+1$.

It follows that the lex-segment ideal with h-vector as above has exactly one generator in degree $d+1$. This means that there is only one monomial T in $\mathcal{G}_{d+1}$ and so

$$\beta_{n-2,d+n-1} = \binom{m(T)-1}{n-2} \leq \binom{n-1}{n-2} = n-1.$$

(In fact, it is either equal to $n-1$ or to 1.)

How many generators does the lex-segment ideal with this h-vector have in degree d? It is easy to see that the number of generators of the lex-segment ideal in degree d is exactly $h_{d-1}^{<d-1>} - p$, so we need to estimate this number.

Claim. $h_{d-1}^{<d-1>} - p \geq n+1$.

Proof of Claim. Note that we are assuming that $h_{d-1} \geq p+n$. Recall also that $\_^{<d-1>}$ is a strictly increasing function, so it is enough to prove the claim for the least possible value of h_{d-1}, namely $p+n$ $(n > 0)$.

Now, since $d+2 \leq p \leq 2d$ we have that the $d-1$ binomial expansion of p is given as follows.

for $p = 2d$, $\quad p = \binom{d}{d-1} + \binom{d-1}{d-2} + \binom{d-3}{d-3};$

for $p = 2d-1$, $\quad p = \binom{d}{d-1} + \binom{d-1}{d-2};$

for $p = d+r,\ 2 \leq r \leq d-2$, $\quad p = \binom{d}{d-1} + \binom{d-2}{d-2} + \cdots + \binom{d-r-1}{d-r-1}.$

Thus,
$$p^{<d-1>} - p = \begin{cases} 2, & \text{if } p = 2d, \\ 2, & \text{if } p = 2d-1, \\ 1, & \text{if } p = d+r, \ 2 \leq r \leq d-2. \end{cases}$$

We can thus conclude that
$$h_{d-1}^{<d-1>} - p \geq (p+n)^{<d-1>} - p \geq p^{<d-1>} - p + n \geq n+1$$
as we claimed.

Now since $p \leq 2d$ and $d \geq n+2$, it is enough to look at the last $3d$ monomials of degree d to see at least $n+1$ of the generators of I in degree d.

The last $3d$ monomials of degree d are
$$\underbrace{x_{n-2}^2 x_{n-1}^{d-2}, x_{n-2}^2 x_{n-1}^{d-3} x_n, \ldots, x_{n-1}^2 x_n^{d-2}}_{d-1}, \underbrace{x_{n-2} x_{n-1}^{d-1}, x_{n-2} x_{n-1}^{d-2} x_n, \ldots, x_{n-2} x_n^{d-1}}_{d},$$
$$\underbrace{x_{n-1}^d, x_{n-1}^{d-1} x_n, \ldots, x_n^d}_{d+1}.$$

Notice that among these, only the monomials T in

(3.4) $$\{x_{n-1}^d, x_{n-2} x_{n-1}^{d-1}, x_{n-2}^2 x_{n-1}^{d-2}\}$$

satisfy $m(T) = n-1$. Moreover, any consecutive collection of monomials among these last $3d$ which contains two elements of (3.4) contains at least d monomials and $d \geq n+2$. Put another way, for any consecutive $n+1$ monomials (among these last $3d$) there are always at least n for which $m(T) = n$.

It follows that
$$\beta_{n-1,n+d-1} = \sum_{T \in \mathcal{G}(I)_d} \binom{m(T)-1}{n-1} \geq n$$
and hence
$$\beta_{n-1,n+d-1} > \beta_{n-2,n+d-1}$$
as we wanted to show. $\square$

Here is another non-cancelation result.

PROPOSITION 3.9. *Let h_{d-2}, h_{d-1}, h_d be three integers such that*
$$h_d = h_{d-1}^{<d-1>} \quad \text{and} \quad h_{d-1} = h_{d-2}^{<d-2>}.$$
Let I be any ideal in $R = k[x_1, \ldots, x_n]$ for which
$$\begin{aligned} \mathbf{H}(R/I, d-2) &= h_{d-2} + \varepsilon, \ \varepsilon \geq n, \\ \mathbf{H}(R/I, d-1) &= h_{d-1}, \\ \mathbf{H}(R/I, d) &= h_d - 1. \end{aligned}$$
Then $\dim_k \operatorname{soc}(R/I)_{d-2} \geq 1$ and so any O-sequence
$$(1, n, \ldots, h_{d-2} + \varepsilon, h_{d-1}, h_d - 1, \ldots, h_s)$$
of length $s+1$ ($s \geq d$) is not a level O-sequence.

PROOF. As we saw in the proof of Theorem 3.4, it is enough to show that when I is a lex-segment Artinian ideal having socle degree d and satisfying the conditions of the Proposition, then $\beta_{n-1,d-2+n}(R/I) > \beta_{n-2,d-2+n}(R/I)$.

So, let's suppose that I is a lex-segment Artinian ideal satisfying the conditions of the Proposition.

Claim. $\dim_k \operatorname{soc}(R/I)_{d-2} = \varepsilon$.

Proof of Claim. Let $V \subset R_{d-2}$ be the lex-segment subspace of codimension h_{d-2}. The assumption that $h_{d-1} = h_{d-2}^{<d-1>}$ implies that $R_1 V = I_{d-1}$. Since $-^{<d-1>}$ is a strictly increasing function we know that

$$V = \{f \in R_{d-2} \mid R_1 f \subseteq I_{d-2}\}.$$

Thus, $V/I_{d-2} = \operatorname{soc}(R/I)_{d-2}$.

The hypothesis gives that the codimension of I_{d-2} in V is exactly ε and that completes the proof of the Claim.

Since we know the dimension of the socle of R/I in degree $d-2$ is ε, we obtain

$$\beta_{n-1,d-2+n}(R/I) = \varepsilon.$$

So, it will be enough to show that $\beta_{n-2,d-2+n}(R/I) < \varepsilon$. But,

$$\beta_{n-2,d-2+n}(R/I) = \sum_{T \in \mathcal{G}(I)_d} \binom{m(T)-1}{n-2}$$

by the Eliahou-Kervaire result. Our assumption is that $\mathbf{H}(R/I, d) = h_d - 1$ and that gives us that I has exactly one generator, T, in degree d. If $m(T) = n$ then $\beta = \beta_{n-2,d-2+n}(R/I) = n-1$; if $m(T) = n-1$ then $\beta = 1$ and if $m(T) < n-1$ then $\beta = 0$. In each case, $\varepsilon > \beta$ and so we are done. □

REMARK 3.10. We can apply this Proposition to the O-sequences $(1, 3, \ldots, \geq 6, 3, 2)$ since 3_{s-2}, 3_{s-1}, 3_s satisfies the maximal growth criterion of the Proposition as soon as $s - 2 \geq 3$.

Similarly, we can apply the proposition to the O-sequence $(1, 3, 6, 10, 9, 10)$ since the triple 7_3, 9_4, 11_5 satisfies the maximal growth criterion.

Although it is important to have simple criteria on an h-vector which allow us to recognize when there is "non-cancelation" (see Remark 3.7) i.e., theorems like Theorem 3.4, Proposition 3.8, and Proposition 3.9, it would be wrong to assume that "non-cancelation" is the only thing that prevents an h-vector from being a level sequence (something which is true for h-vectors which begin $(1, 2, \ldots)$, see [24]).

The following examples (and several more which we will indicate in the Appendix) show that even when an h-vector seems to permit cancelation arithmetically, that cancelation need not be possible.

EXAMPLE 3.11. Consider the h-vector $h = (1, 3, 5, 7, 6, 6, 2)$. The Betti diagram of the lex-segment ideal with this h-vector is

$$
\begin{array}{c|ccc}
 & 0 & 1 & 2 \\ \hline
0 & 1 & - & - \\
1 & - & 1 & - \\
2 & - & - & - \\
3 & - & 3 & 5 \\
4 & - & 1 & 2 \\
5 & - & 5 & 9 \\
6 & - & 2 & 4 \\
\end{array}
\begin{array}{c}
- \\ - \\ - \\ 2 \\ 1 \\ 4 \\ 2 \\
\end{array}
$$

Now suppose that $A = R/I$, $R = k[x_1, x_2, x_3]$ is a level algebra with h-vector h as above.

Claim. The minimal number of generators of I, in degree 6, is < 5.

Notice that once this claim is proved we are done. Since then $\beta_{0,6}(A) < 5$ and by the Bigatti-Hulett-Pardue result (see (3.1) above) we must have $\beta_{1,6}(A) < 2$. This, in turn, implies that $\beta_{2,6}(A) \geq 1$ and hence A cannot be level.

Proof of Claim. Suppose that I has 5 generators in degree 6 and let $J = I_{\leq 5}$. Then the Hilbert function of R/J begins 1 3 5 7 6 6 7 $\cdots$. Now, we can use Theorem 3.4 to assert that R/J has 1-dimensional socle in degree 4. Since R/J and R/I agree in degree ≤ 5 we conclude that R/I cannot be level.

We note that, using the same kind of argument, one can show that the following h-vectors are also not level sequences:

$$(1,3,6,7,6,6,2), \quad (1,3,5,7,5,4,2), \quad (1,3,6,8,6,5,2),$$
$$(1,3,6,9,7,6,2), \quad (1,3,6,10,6,4,2), \quad (1,3,6,10,7,6,2).$$

EXAMPLE 3.12. To show just how far from "cancelation" we can be without obtaining a level sequence, we consider another example. Let $h = (1, 3, 6, 9, 6, 4, 2)$. The Betti diagram for the lex-segment ideal with this h-vector is

$$
\begin{array}{c|ccc}
 & 0 & 1 & 2 \\ \hline
0 & 1 & - & - \\
1 & - & - & - \\
2 & - & 1 & - \\
3 & - & 6 & 10 \\
4 & - & 3 & 5 \\
5 & - & 2 & 4 \\
6 & - & 2 & 4 \\
\end{array}
\begin{array}{c}
- \\ - \\ - \\ 4 \\ 2 \\ 2 \\ 2 \\
\end{array}
$$

If there were a level algebra with this h-vector we'd have to be able to cancel $\beta_{2,6} = 4$ from this diagram.

So, suppose that $A = R/I$ is a level algebra with this h-vector. If I has no generators in degree 6 then we'd have a contradiction: for then $\beta_{0,6} = 0$ and so $\beta_{1,6}(A) \leq 3$. But, we need $\beta_{1,6}(A) \geq 4$ to cancel the $\beta_{2,6} = 4$.

So, it suffices to show that I has no generators in degree 6 (we know that it has ≤ 2 generators in degree 6).

Case 1. Suppose I has two generators in degree 6. Let $J = \langle I_{\leq 5} \rangle$. Then the Hilbert function of R/J begins 1 3 6 9 6 4 4 $\cdots$. By Theorem 3.4, R/J has a 2-dimensional socle in degree 4 and hence so does I. That is a contradiction.

Case 2. Now suppose that I has one generator in degree 6 and let $J = \langle I_{\leq 5} \rangle$ as above. Then the Hilbert function of $B = R/J$ begins 1 3 6 9 6 4 3 $t \cdots$. We don't know exactly what t is but we can say that $0 \leq t \leq 3$.

Let's first consider the possibility that $t = 3$. Since $J \subseteq I$ we have a canonical surjection R/J onto R/I, which is an isomorphism in degree ≤ 5. By Theorem 3.4, R/J has non-zero socle in degree 5. The surjection (which is an isomorphism in degree 5) carries that non-zero socle into non-zero socle of R/I in degree 5, which is a contradiction.

So, assume that $t \leq 2$. The lex-segment ideal whose h-vector is $h = (1, 3, 6, 9, 6, 4, 3, t)$ has Betti diagram which starts

$$\begin{array}{c|cccc} & 0 & 1 & 2 \\ \hline 0 & 1 & - & - \\ 1 & - & - & - \\ 2 & - & 1 & - \\ 3 & - & 6 & 10 & 4 \\ 4 & - & 3 & 5 & 2 \\ 5 & - & 1 & 2 & 1 \\ 6 & - & 3-t & 6-2t & 3-t \\ 7 & & & & \end{array}$$

Since A is level, $\beta_{2,7}(A) = 0$ and since A and B agree in degree less than or equal to 5, that implies that B has no socle in degree 4 either, so $\beta_{2,7}(B) = 0$ as well. That implies that $\beta_{1,7}(B) = 0$ as well. This, in turn, implies that $\beta_{0,7}(B) = 3 - t$ exactly. But $3 - t > 0$ and so J has generators in degree 7, which is a contradiction.

In an entirely similar way we can show that

$$(1, 3, 6, 10, 7, 5, 2) \text{ and } (1, 3, 5, 7, 9, 5, 2)$$

are not level sequences.

EXAMPLE 3.13. In this example we show how the methods of this section can be supplemented by other arguments to give even more subtle non-existence results.

Consider the h-vector $h = (1, 3, 6, 5, 4, 3)$. The corresponding lexsegment ideal has this Betti diagram:

(3.5)
$$\begin{array}{c|cccc} & 0 & 1 & 2 \\ \hline 0 & 1 & - & - & - \\ 1 & - & - & - & - \\ 2 & - & 5 & 6 & 2 \\ 3 & - & 2 & 3 & 1 \\ 4 & - & 1 & 2 & 1 \\ 5 & - & 3 & 6 & 3 \end{array}$$

Suppose that h is a level sequence, and let R/I be a level algebra with this Hilbert function. We proceed step by step.

Note that I has either 1 or 0 minimal generators of degree 5. If I has one minimal generator in degree 5, then the "usual" argument works: let J be the ideal generated by $\langle I_{\leq 4} \rangle$. Then R/J has Hilbert function $(1, 3, 6, 5, 4, 4, 4, \ldots)$, hence has socle in degree 3 thanks to Proposition 3.8 (b). Thus R/I also has socle in degree

3. Therefore without loss of generality we can assume that there are no minimal generators in degree 5.

Notice also that I has either 0,1 or 2 generators in degree 4, and either 3 or 2 generators in degree 6, thanks to cancelation.

Let's rewrite the Betti diagram for R/I after doing all the canceling we can (to make it level):

$$\begin{pmatrix} & 0 & 1 & 2 \\ 0 & 1 & - & - & - \\ 1 & - & - & - & - \\ 2 & - & 5 & 4+a & - \\ 3 & - & a & - & - \\ 4 & - & - & b & - \\ 5 & - & 2+b & 5 & 3 \end{pmatrix} \qquad (a = 0, 1 \text{ or } 2, \ b = 0 \text{ or } 1)$$

If $b = 1$ then I has 3 minimal generators of degree 6. As before we take $J = \langle I_{\leq 5} \rangle$. R/J has Hilbert function $(1, 3, 6, 5, 4, 3, 3, 3, \ldots)$ so our Proposition 3.8 (or the result of Cho-Iarrobino, Theorem 3.4) says that R/J has socle in degree 4, hence so does R/I. So $b = 0$.

If $a = 2$ then I has 2 minimal generators of degree 4. Take $J = \langle I_{\leq 3} \rangle$. R/J has Hilbert function $(1, 3, 6, 5, 6, 7, \ldots)$. By the result of Cho and Iarrobino (Theorem 3.4), R/J has socle in degree 2, so also does R/I. So we eliminate $a = 2$.

We now have the diagram

$$\begin{pmatrix} & 0 & 1 & 2 \\ 0 & 1 & - & - & - \\ 1 & - & - & - & - \\ 2 & - & 5 & 4+a & - \\ 3 & - & a & - & - \\ 4 & - & - & - & - \\ 5 & - & 2 & 5 & 3 \end{pmatrix} \qquad (a = 0 \text{ or } 1)$$

We now ask what kind of regular sequence we can find in I.

Step 1. Note that $\dim I_3 = 5$. Suppose that everything in I_3 has a common factor. If this factor were quadratic, say Q, it would not be possible to find five independent polynomials of the form $L_i Q$ where L_i is linear. So the common factor must be linear. Then I_3 has the form $\langle LQ_1, \ldots, LQ_5 \rangle$. Let us compute the dimension of this in degree 4:

$$\dim \langle LQ_1, \ldots, LQ_5 \rangle_4 = \dim L \cdot \langle Q_1, \ldots, Q_5 \rangle_3 \leq 10$$

since the vector space of *all* cubics has dimension 10. But I_4 has at most one generator of degree 4, and $\dim I_4 = 11$, so in fact $a = 1$ and $\dim \langle LQ_1, \ldots, LQ_5 \rangle_4 = 10$ and $\langle Q_1, \ldots, Q_5 \rangle_3$ is all of R_3. On the other hand, let Q be an element of R_2 that is not in $\langle Q_1, \ldots, Q_5 \rangle$. Consider the element LQ. It is not zero in R/I, but one can see that it is a socle element. This shows that in I_3 we at least have a regular sequence of length 2.

Step 2. We check that I_3 does not have a regular sequence of length 3. Indeed, if it did we could link I with such a regular sequence, and one quickly checks that the residual would have to have Hilbert function $(1, 0, 2, 2)$ which is obviously impossible.

Step 3. We check that I does not have a regular sequence of type $(3, 3, 4)$. If it did, linking would give an algebra with Hilbert function $(1, 3, 3, 4, 3)$. Note that the growth from degree 2 to degree 3 is maximal, so the cubics have a GCD of degree 1. Therefore such an ideal does not have a regular sequence of two cubics. Contradiction.

Step 4. Since the only other generators of I come in degree 6, we must have a regular sequence of type $(3, 3, 6)$. Linking gives a residual, J, with Hilbert function $(1, 3, 6, 8, 6, 5, 3)$. Let us consider the minimal free resolution of the residual. Note first that since we know (almost) the Betti diagram of I, the usual liaison tricks give that J has Betti diagram

$$
\begin{array}{c|cccc}
 & 0 & 1 & 2 \\
\hline
0 & 1 & - & - & - \\
1 & - & - & - & - \\
2 & - & 2 & - & - \\
3 & - & 3 & 5 & 1 \\
4 & - & - & - & - \\
5 & - & 1 & - & a \\
6 & - & - & 4+a & 3 \\
\end{array}
\qquad (a = 0 \text{ or } 1)
$$

We know that the two cubic generators form a regular sequence. If J contained a regular sequence of type $(3, 3, 4)$, the residual would have Hilbert function $(1, 0, 1, 2)$, which is impossible. So the smallest regular sequence is still $(3, 3, 6)$.

Let $J' = \langle J_{\leq 5} \rangle$ (i.e. remove the generator of degree 6). R/J' has Hilbert function $(1, 3, 6, 8, 6, 5, 4, \ldots)$ and R/J' has Krull dimension 1, since the longest regular sequence has length 2. That is, the saturation of J' defines a zeroscheme, Z, in $\mathbb{P}^2$. Note that since $\dim(R/J')_6 = 4$, we get $\deg Z \leq 4$.

Step 5. Now, the two cubics give the saturated ideal of a degree 9 zeroscheme (a complete intersection) that contains Z. We add three forms of degree 4 to get J'. Let us consider these three forms one at a time.

Let us call the two cubics G_1 and G_2 (they form a regular sequence) and let us call the quartics F_1, F_2 and F_3. The Hilbert function of $R/(G_1, G_2)$ is

$$(1, 3, 6, 8, 9, 9, \ldots).$$

Now we add F_1, which we can assume without loss of generality is chosen generally in J'_4. We note (using liaison):

- F_1 contains at most 4 of the 9 complete intersection points defined by G_1, G_2. If L is a linear form, it is impossible for LF_1 to vanish on all 9 points. Therefore $H(R/(G_1, G_2, F_1), 5) = 9 - 3 = 6$.

- If Q is a quadratic form, then QF_1 vanishes at all 9 points of the complete intersection if and only if Q vanishes at the residual to Z in this complete intersection. Because Z consists of at most 4 points, this residual can lie on at most one conic. Therefore $H(R/(G_1, G_2, F_1), 6)$ is either $9 - 6 = 3$ (if the residual lies on no conic) or $9 - 5 = 4$ (if the residual lies on one conic). But $\dim(R/J')_6 = 4$, so in fact it must be the second possibility.

So we have the Hilbert function of $R/(G_1, G_2, F_1)$:

$$(1, 3, 6, 8, 8, 6, 4, \ldots).$$

Now we add F_2, noting that the value of the Hilbert function in degree 6 is already at the desired value! What happens in degrees 4 and 5? Degree 4 clearly becomes 7. In degree 5, the Hilbert function of R/J' is 5. So if we choose F_2 also generally in J', the value of the Hilbert function of $R/(G_1, G_2, F_1, F_2)$ in degree 5 has to drop from 6 to 5 (since otherwise it must remain 6 even when we add F_3). So the Hilbert function of $R/(G_1, G_2, F_1, F_2)$ is

$$(1, 3, 6, 8, 7, 5, 4, \ldots).$$

Now recall that the Hilbert function of R/J' is $(1, 3, 6, 8, 6, 5, 4, \ldots)$. Therefore, the third generator F_3 must be a socle element of $R/(G_1, G_2, F_1, F_2)$.

The original ideal J has generators $G_1, G_2, F_1, F_2, F_3, A$ where A is a generator of degree 6. The Hilbert function of R/J is $(1, 3, 6, 8, 6, 5, 3, 0)$. Consider the ideal $J'' = (G_1, G_2, F_1, F_2, A)$. This is a quotient of $R/(G_1, G_2, F_1, F_2)$, which we recall has Hilbert function $(1, 3, 6, 8, 7, 5, 4, \ldots)$.

Since F_3 is a socle element, as described above, the fact that $G_1, G_2, F_1, F_2, F_3, A$ span all of R_7 implies that G_1, G_2, F_1, F_2, A also span all of R_7 (since xF_3, yF_3 and zF_3 are all in the ideal (G_1, G_2, F_1, F_2)). So the Hilbert function of the quotient $R/(G_1, G_2, F_1, F_2, A)$ is $(1, 3, 6, 8, 7, 5, 3, 0)$. Now, (G_1, G_2, A) is a regular sequence, so we can link. The residual has Hilbert function $(1, 3, 6, 5, 4, 2, 0)$.

The generators of J'' have degrees 3,3,4,4,6. The complete intersection uses the generators of degrees 3,3,6. Therefore, using the usual mapping cone, the residual is level! But we will see in Example 5.7 that $(1, 3, 6, 5, 4, 2)$ is not a level sequence. So this contradiction shows (finally) that $(1, 3, 6, 5, 4, 3)$ is not a level sequence.

CHAPTER 4

Some Refinements

Until this point we have focused on results that can be used to show that sequences are not level. In the coming chapters, we will be more interested in showing that sequences *are* level. This will be done by construction methods, and by means of results that obtain new level sequences from previously known ones. However, we will also want to refine our search somewhat. We will want to know not only which sequences are level, but in fact which sequences exist for level algebras with certain additional properties. In this short chapter we describe these properties.

First we extend our idea of a level algebra beyond the Artinian horizon! We take our clue from the fact that there is a strong relationship between a level Artinian algebra and properties of its minimal free resolution.

Let A be any standard graded k-algebra. Recall that A is called *Cohen-Macaulay* (C-M) if the length of the longest regular sequence in A is the Krull dimension of A.

Alternatively, given A, write $A = P/I$ where $P = k[x_0, \ldots, x_n]$. Suppose that the Krull dimension of A is d. Let $\mathbb{F}$ be a minimal graded free resolution of A as P-module and write:

$$\mathbb{F}: \quad 0 \to \mathcal{F}_\ell \to \cdots \to \mathcal{F}_0 \to P \to A \to 0$$

where we know (Hilbert) that $\ell \leq n$.

It is a well-known theorem of Auslander and Buchsbaum that: A is Cohen-Macaulay if and only if $\ell = n - d$.

Now, if k is an infinite field and A is a C-M k-algebra of Krull dimension d then one can always find a maximal regular sequence in A consisting of forms in A having degree 1. If we let $\overline{L_1}, \ldots, \overline{L_d}$ be such a regular sequence in A then $B = A/(\overline{L_1}, \ldots, \overline{L_d})$ is an Artinian algebra. (The process of taking the quotient of a C-M k-algebra by an ideal generated by a maximal regular sequence of linear forms, is often referred to as *Artinian reduction*.)

We can rewrite B as $B = P/(I, L_1, \ldots, L_d) \simeq k[x_0, \ldots, x_{n-d}]/J$. If we let $\tilde{P} = k[x_0, \ldots, x_{n-d}]$, then B has minimal free resolution (as a $\tilde{P}$-module)

$$\mathbb{H}: \quad 0 \to \mathcal{H}_{n-d} \to \cdots \to \mathcal{H}_0 \to \tilde{P} \to B \to 0.$$

Moreover, the ith graded Betti numbers of A (as a P-module) are the same as the ith graded Betti numbers of B (as a $\tilde{P}$-module).

With these observations out of the way we now make our general definition of a level k-algebra.

DEFINITION 4.1. Let A be a standard graded Cohen-Macaulay k-algebra of Krull dimension d and write $A = P/I$, $P = k[x_0, \ldots, x_n]$ and $\mathbf{H}(A, 1) = n+1$. Let

$$\mathbb{F}: \quad 0 \to \mathcal{F}_{n-d} \to \cdots \to \mathcal{F}_1 \to \mathcal{F}_0 \to P \to A \to 0$$

be a minimal graded free resolution of A, as a graded P-module.

A is a *level algebra* if and only if $\mathcal{F}_{n-d} = P(-a)^b$ for some positive integers a and b.

Now we would like to consider a different refinement of the notion of a level algebra. First some definitions.

DEFINITION 4.2. 1) An Artinian k-algebra $A = \oplus_{i=0}^{s} A_i$ ($A_s \neq 0$), is said to satisfy the *Weak Lefschetz Property* (WLP for short) if there is an element $L \in A_1$ (called a *Lefschetz element*) such that the linear transformations

$$L : A_i \to A_{i+1}, \quad 1 \leq i \leq s-1$$

(given by multiplication by L) are either injective or surjective, i.e., have maximal rank, for each i.

2) An O-sequence $(1, h_1, \ldots, h_s)$, with $h_s \neq 0$ is called *unimodal* if there is an integer t such that

$$h_1 \leq \ldots \leq h_t \geq h_{t+1} \geq \ldots \geq h_s.$$

REMARK 4.3. Notice that if A has the WLP and L is a Lefschetz element such that the multiplication map $\times L : A_i \to A_{i+1}$ is surjective then, for any $r \geq 1$, the multiplication map $\times L : A_{i+r} \to A_{i+1+r}$ is also surjective. This follows immediately from the fact that A is a standard algebra. As a consequence we see that if A is an Artinian algebra with the WLP then the h-vector of A is unimodal.

But, the WLP is a strictly stronger condition: an Artinian standard graded algebra can have the unimodal vector $h = (1, 3, 6, 6, 7, 2)$ as its h-vector but such an algebra could never, by what we just observed, satisfy the WLP.

As mentioned in the introduction, R. Stanley is the "godfather" of this concept and studied it implicitly in several of his papers ([**66**], [**67**], [**68**]). Since then, this concept has received a great deal of attention in the study of Artinian Gorenstein algebras. For example, there is a *complete* characterization of the h-vectors of Gorenstein Artinian algebras with the WLP [**32**] even though we only have a complete characterization of all Gorenstein h-vectors in the case of codimension ≤ 3.

Moreover, it is an open question as to whether **ALL** Gorenstein Artinian quotients of $k[x, y, z]$ have the WLP (it is known that all complete intersection quotients of $k[x, y, z]$ satisfy the WLP – see [**32**], [**35**], [**54**], [**55**], and [**72**] for this and related interesting results).

We will see, in the next chapter, that there are many ways to construct Artinian level algebras with the WLP. In fact, we wonder whether (at least in codimension 3) all the h-vectors of Artinian level algebras are the h-vectors of level algebras with the WLP. All our investigations seem to indicate that this is indeed the case.

Indeed, one of the motivating questions behind this work is whether for every level sequence $h = (h_i)$ ($0 \leq i \leq s$), there in fact exists a level Artinian algebra *with the WLP* whose Hilbert function is h. Consider the sequence

$$h' = (h'_i) \text{ where } h'_i = \max\{h_i - h_{i-1}, 0\}$$

Recall from [**35**] the fact that if A has the WLP then h' must again be an O-sequence (equivalently, if t is the last index for which $h_t \geq h_i$ for every i, then $(1, h_1, \ldots, h_t)$ is a *differentiable* O-sequence). Furthermore h must be unimodal. When the number of variables is five or more, it is known that the Hilbert function of Artinian Gorenstein algebras need not be unimodal ([**3**], [**10**], [**11**]), so level algebras need not have WLP. But in four or fewer variables, this is open.

We would now like to pose some questions that are directly (or indirectly) related to the WLP for level algebras.

QUESTION 4.4. Let $h = (1, n, h_2, \ldots, h_s)$ be the h-vector of a level algebra.
1) If $n = 3$ or $n = 4$, is it true that h is also the h-vector of a level algebra with the WLP?
2) If $n = 3$, is it possible that *every* level algebra has the WLP? Note that Ikeda ([**47**]) and Boij ([**9**]) gave examples of Gorenstein algebras that do not have the WLP when $n = 4$.
3) If Question 1) has a negative answer, and if $n = 3$ or $n = 4$, must h at least be unimodal?
4) If $n = 3, 4$, and if t is the last index for which $h_t \geq h_i$ for every i, is it true that $(1, n, h_2, \ldots, h_t)$ is a differentiable O-sequence? (For $n = 3$, this is the case for every example in the appendix.)

These questions are studied further in the appendix.

CHAPTER 5

Constructing Artinian Level Algebras

In Chapters 2 and 3 we saw that there were conditions on the h-vector of an Artinian graded algebra that would prohibit it from being a level O-sequence. In this chapter and the next we take a different point of view and show how one can construct level algebras, i.e., show that certain h-vectors are level O-sequences.

Our constructions, broadly speaking, move in two different directions. With the first construction methods, in this chapter, we find Artinian level algebras. The other constructions, in the following chapter, aim at finding higher dimensional objects that will give us Artinian level algebras. In both chapters, some of our results produce level sequences (and algebras) from scratch, while others deduce the existence of level sequences (and algebras) from previously known ones.

5.1. Inverse Systems

We now recall a very interesting method for constructing Artinian level algebras. This method is based on the idea of *Macaulay's Inverse Systems*. We will only give a quick review of the method and refer the reader to either [21] or [46] for more details.

Let $R = k[x_1, \ldots, x_n]$ and $S = k[y_1, \ldots, y_n]$. We can consider S as a graded R-module by: if $F \in S_j$ then $x_i \circ F = (\frac{\partial}{\partial y_i})F$. We extend this action in the obvious way and note that the action *lowers* degree on S and hence S is not a finitely generated R-module.

There is an order reversing function from the ideals of R to the R-submodules of S defined by:

$$\varphi_1 : \{\text{ideals of } R\} \to \{R\text{-submodules of } S\}$$

where

$$\varphi_1(I) = \{F \in S \mid G \circ F = 0 \text{ for all } G \in I\}$$

This is a 1-1 correspondence whose inverse (φ_2) is given by $\varphi_2(M) = \mathrm{ann}_R(M)$. In fact, we denote $\varphi_1(I)$ by I^{-1}, which is called the *inverse system* to I.

It is very easy to construct I^{-1} (and this is at the heart of the proof of the 1-1 correspondence). One first observes that the pairing

$$R_j \times S_j \longrightarrow S_0 \simeq k$$

is a perfect pairing and so S_j can be identified with R_j^* (the dual vector space to R_j). If V is a subspace of R_j we write $V^\perp$ for the annihilator of V in this pairing. Then, if $I \subset R$ is an ideal and I_j its jth graded piece, then Macaulay observed that:

$$(I^{-1})_j = I_j^\perp.$$

It follows immediately that

$$\dim_k (I^{-1})_j = \dim_k R_j - \dim_k I_j = \mathbf{H}(R/I, j) .$$

It is a simple consequence of this last observation that I^{-1} is a finitely generated R-submodule of S if and only if R/I is Artinian.

REMARK 5.1. There is another way to define Inverse Systems which considers S as an R-module in a different way. In this other method, we consider the *contraction* operations, D_{x_i} where, if F is a monomial in S_j then

$$D_{x_i}(F) = \begin{cases} 0, & \text{if } y_i \text{ does not divide } F, \\ F/y_i & \text{if } y_i \text{ divides } F. \end{cases}$$

We extend this action to all of S in the obvious way and recall that when the characteristic of k is 0, this action is equivalent to the one described above. The contraction operation has the advantage that it doesn't end up increasing the sizes of coefficients. (See [21] or [46] for more details.)

The really interesting connection between inverse systems and what we've been considering is the following theorem of Macaulay. We continue with the notation above.

THEOREM 5.2 (Macaulay). *Let I be an Artinian ideal of R and I^{-1} its inverse system. Then I^{-1} has exactly ν_j minimal generators of degree j if and only if the socle of R/I in degree j has dimension exactly ν_j.*

REMARK 5.3. 1) This gives us a new interpretation of the socle vector of an Artinian algebra of the form $A = k[x_1, \ldots, x_n]/I$. The entries of the socle vector tell us the number of generators of the inverse system of I in each degree.

2) Since we are interested in level algebras (Artinian, say, with socle degree s, type c and embedding dimension n) then this theorem tells us how to make **all** of them. We look at every subspace of $S_s = k[y_1, \ldots, y_n]_s$ of dimension c and form the R-submodule of S generated by that subspace. The result is a level algebra of the type we are looking for and every level algebra of socle degree s, type c and embedding dimension n arises in this way.

E.g., suppose we would like to construct a level algebra with socle degree 4, embedding dimension 3 and type 2. Macaulay's Theorem says we have to look at a two dimensional vector space of S_4, where $S = k[y_1, y_2, y_3]$ and take the inverse system it generates.

For example, consider the vector space of S_4 generated by $F_1 = y_1^4$ and $F_2 = y_2^4 + y_3^4$. The inverse system, call it M generated by these two elements of degree 4 will have $M_3 = \langle y_1^3, y_2^3, y_3^3 \rangle$, $M_2 = \langle y_1^2, y_2^2, y_3^2 \rangle$, $M_1 = \langle y_1, y_2, y_3 \rangle$, and $M_0 = \langle 1 \rangle$. So, if $I = \operatorname{ann}_R(M)$ and $A = k[x_1, x_2, x_3]/I$ then the h-vector of A is $(1, 3, 3, 3, 2)$.

3) We find it very interesting that there is this strong connection between the dimensions of spaces of partial derivatives of a collection of forms (of the same degree) and our study of level algebras. E.g., apart from an appeal to Theorem 2.2 iii), we know of no other way to prove that if F is a form of degree s then the dimension of the space of its ith partial derivatives and the dimension of its space of $(s-i)^{th}$ partial derivatives are equal.

The "inverse system" point of view on level algebras has some simple and useful consequences.

5.1. INVERSE SYSTEMS

PROPOSITION 5.4. *Let $(1, n, h_2, \ldots, h_s)$ be the h-vector of an Artinian level algebra with socle degree s. Then*

$$(5.1) \qquad h_i \leq \min\left\{\binom{n-1+i}{n-1}, h_s\binom{n-1+s-i}{n-1}\right\}.$$

PROOF. (See also Lemma 2.7.) Let $R = k[x_1, \ldots, x_n]$ and suppose $A = R/I$ is an Artinian level algebra with the given h-vector. That $h_i \leq \dim_k R_i = \binom{n-1+i}{n-1}$ is obvious.

Let $F_1, \ldots, F_{h_s} \in S_s$ be the generators of the inverse system of I. Each of these forms has at most $\binom{n-1+s-i}{n-1}$, independent $(s-i)^{th}$ partial derivatives and so h_s forms cannot have their $(s-i)^{th}$ partial derivatives generate a space of dimension greater than $h_s\binom{n-1+s-i}{n-1}$. □

Artin level algebras for which the inequality above is an equality, are what Iarrobino [43] (see also [20]) calls *compressed level algebras*. He showed that compressed level algebras always exist (see also [20]). Roughly speaking the result comes from the fact that a **general** set of h_s forms of degree s in P will generate an inverse system of an ideal I for which the h-vector of $A = P/I$ satisfies the equalities of (5.1) for every $i \leq s$.

Inverse systems can also be used to produce new level algebras from known level algebras. Before stating the Proposition we want to recall some beautiful results from [43].

Let R and S be as above and let $L_1, \ldots, L_t$ be general linear forms in S_1. Fix a positive integer d and suppose that $t < \binom{d+n-1}{d}$. Define

$$F_d = L_1^d + \cdots + L_t^d$$

and let I^{-1} be the R-submodule of S generated by F.

Then, Iarrobino shows in [43] that the h-vector of the Gorenstein Artinian algebra R/I is $\mathbf{H}(d, t, n)$, where the ith entry of $\mathbf{H}(d, t, n)$ is given by:

$$\mathbf{H}(d,t,n)_i = \min\left\{t, \binom{i+n-1}{i}, \binom{d-i+n-1}{d-i}\right\}.$$

Actually, Iarrobino shows an even stronger result. Without going into the details of the proof, he shows that given a subspace U of S_d and the R-submodule of S generated by U then, with certain natural restrictions on t, there are general linear forms $L_1, \ldots, L_t$ so that the R-submodule of S generated by F (as above) is as disjoint as possible from the R-submodule of S generated by U.

The application to level algebras is

PROPOSITION 5.5 ([43], Theorem 4.8A). *Let $h = (1, h_1, \ldots, h_s)$ be the h-vector of a level algebra $A = R/I$ where $R = k[x_1, \ldots, x_n]$. Let $t < \binom{s+n-1}{s}$ and define the sequence $h + \mathbf{H}(s, t, n)$ as follows: the ith entry of this sequence is*

$$(h + \mathbf{H}(s,t,n))_i := \min\left\{h_i + \mathbf{H}(s,t,n)_i, \binom{i+n-1}{i}\right\}.$$

Then $h + \mathbf{H}(s, t, n)$ is also the h-vector of a level algebra.

REMARK 5.6. This Proposition is very useful in showing the existence of level algebras. E.g., we know, (see [64]) that there are Gorenstein Artinian algebras with socle degree 6 and embedding dimension ≤ 3 having h-vector:

$(1,2,2,2,2,2,1)$, $(1,2,3,3,3,2,1)$, $(1,2,3,4,3,2,1)$, $(1,3,3,3,3,3,1)$,
$(1,3,4,4,4,3,1)$, $(1,3,4,5,4,3,1)$, $(1,3,5,5,5,3,1)$, $(1,3,5,6,5,3,1)$,
$(1,3,6,6,6,3,1)$, $(1,3,6,7,6,3,1)$, $(1,3,6,8,6,3,1)$, $(1,3,6,9,6,3,1)$,
$(1,3,6,10,6,3,1)$.

Using Proposition 5.5, with $t = 1$, $n = 3$ and $s = 6$, we obtain immediately that the following are the h-vectors of level algebras of embedding dimension 3 and having socle degree 6 and type 2:

$(1,3,3,3,3,3,2)$, $(1,3,4,4,4,3,2)$, $(1,3,4,5,4,3,2)$, $(1,3,4,4,4,4,2)$,
$(1,3,5,5,5,4,2)$, $(1,3,5,6,5,4,2)$, $(1,3,6,6,6,4,2)$, $(1,3,6,7,6,4,2)$,
$(1,3,6,7,7,4,2)$, $(1,3,6,8,7,4,2)$, $(1,3,6,9,7,4,2)$, $(1,3,6,10,7,4,2)$.

(Note that two Gorenstein sequences give the same level sequence.)

Of course, one could repeat this procedure to obtain other level sequences (of type 3, for example) and so on.

One can also use the inverse system approach to show that certain O-sequences are not level sequences.

EXAMPLE 5.7. (See Remark 2.12, c).) Any O-sequence of length $s+1$ and of the form $h = (1, 3, 6, \ldots, 5, 4, 2)$ is not level.

We will apply Corollary 2.11 to this example. First notice that there are only two ways to write the reverse of h as the sum of a Gorenstein sequence of length $s+1$ and an O-sequence by Corollary 2.11. They are:

$$\begin{array}{ccccccc} 1 & 1 & 1 & \cdots & 1 & 1 & 1 \\ 0 & 2 & 5 & \cdots & 4 & 3 & 1 \end{array} \quad \text{and}$$

$$\begin{array}{ccccccc} 1 & 2 & 2 & \cdots & 2 & 2 & 1 \\ 0 & 1 & 4 & \cdots & 3 & 2 & 1. \end{array}$$

But, if there were a level algebra A allowing the first decomposition then it would contain an ideal I for which $I_2 \neq 0$ and $I_{s-2} = 0$. This would imply that A had socle in degree $< s-2$, which is impossible. Thus, we only need consider the second decomposition.

So, suppose there is an ideal $I \subset k[x_1, x_2, x_3] = R$ for which $A = R/I$ is level with h-vector h above. Then $I = \text{ann}(J)$, $J \subset k[y_1, y_2, y_3] = S$, where $J = (F, G)$ is an R-submodule of S generated by two forms of degree s in S.

Moreover, we can assume, with no loss of generality, that the inverse system generated by F gives us the Gorenstein sequence

(5.2) $$1\ 2\ 2\ \cdots\ 2\ 2\ 1.$$

The forms of degree s which generate a submodule of S whose Hilbert function is (5.2) above are precisely the points of the secant variety to the rational normal curve in $\mathbb{P}(S_s)$. A point lies on the secant variety if and only if it lies on a 'true' secant line to the rational normal curve (in which case we can write $F = y_1^s + y_2^s$) or it does not lie on a secant line but rather only on a tangent line to the rational normal curve (in which case we can write $F = y_1^{s-1} y_2$). So, without loss of generality, we can assume that $F = y_1^s + y_2^s$ or $F = y_1^{s-1} y_2$. (For details see either [**21**] or [**46**].)

One easily sees that the inverse system generated by G gives a Gorenstein quotient of R whose Hilbert function is one of the following:

$$1 \quad 3 \quad 4 \quad \cdots \quad 4 \quad 3 \quad 1 \quad \text{or} \quad 1 \quad 3 \quad 5 \quad \cdots \quad 5 \quad 3 \quad 1.$$

Case 1. Suppose that $F = y_1^s + y_2^s$ and suppose that G gives the first of the last two sequences.

Since $h_{s-1} = 4$ we have $\langle G \rangle_{s-1} \cap \langle F \rangle_{s-1} \neq \varnothing$. Let $\alpha y_1^{s-1} + \beta y_2^{s-1} \in \langle G \rangle_{s-1}$ where $(\alpha, \beta) \neq (0,0)$. This gives that either y_1^{s-2} or y_2^{s-2} (or both) are in $\langle G \rangle_{s-2}$. Since $\langle F \rangle_2 = \langle y_1^2, y_2^2 \rangle$ and $\dim_k \langle G \rangle_2 = 4$ and either y_1^2 or y_2^2 is in $\langle G \rangle_2$ we get that

$$\dim_k (\langle F, G \rangle_2) < 2 + 4 = 6 = h_2.$$

This is a contradiction.

So, the only remaining possibility is that the inverse system generated by G gives an Artinian quotient of R with Hilbert function

$$1 \quad 3 \quad 5 \quad \cdots \quad 5 \quad 3 \quad 1.$$

But since $h_{s-2} = 5$, we have that $\langle G \rangle_{s-2} \supseteq \langle F \rangle_{s-2}$ and so $\langle F, G \rangle_i = \langle G \rangle_i$ for all $i \leq s - 2$. But, $h_2 = 6$ and $\dim_k \langle G \rangle_2 = 5$, which is a contradiction.

Case 2. Now assume that $F = y_1^{s-1} y_2$. As in Case 1, if we assume G gives the sequence $1\ 3\ 4\ \cdots\ 4\ 3\ 1$ we must have $\langle F \rangle_{s-1} \cap \langle G \rangle_{s-1} \neq \varnothing$. Assume $\alpha y_1^{s-1} + \beta y_1^{s-2} y_2 \in \langle G \rangle_{s-1}$ with $(\alpha, \beta) \neq (0,0)$. If $\beta = 0$ then $y_1^{s-1} \in \langle G \rangle_{s-1}$. If $\beta \neq 0$ then $y_1^{s-2} \in \langle G \rangle_{s-2}$. In either case we get $\dim_k (\langle F, G \rangle_2) < 4 + 2 = 6 = h_2$, as above.

The argument that G cannot give the sequence $1\ 3\ 5\ \cdots\ 5\ 3\ 1$ is the same as in Case 1 and so is omitted. This completes the argument.

REMARK 5.8. A. Iarrobino has informed us that in his new paper [45], there is a new method introduced which would also eliminate the case $(1, 2, 2, \ldots, 2, 2, 1)$ in Example 5.7 above.

In a similar fashion one can also show that any O-sequence $(1, 3, 5, \ldots, 7, 4, 2)$ or $(1, 3, 5, \ldots, 4, 3, 2)$ is not a level sequence.

A more subtle use of inverse systems in showing non-existence occurs in the following example.

EXAMPLE 5.9. The h-vector $\mathbf{H} = (1, 3, 4, 5, 6, 4, 2)$ is not a level h-vector.

If it were, let I be an Artinian ideal in $R = k[x, y, z]$ so that the Hilbert function of $A = R/I$ is

$$1 \quad 3 \quad 4 \quad 5 \quad 6 \quad 4 \quad 2 \quad 0 \quad \rightarrow .$$

Then we observe, from the Hilbert function, that the forms in I_2 have to have a common factor and that I has no generators in degrees 3 and 4. I.e., we have that $(I_{\leq 4}) = (x^2, xy)$ or (xy, xz). Since both cases can be treated in the same way, we will only deal with the first case here.

Let $S = k[X, Y, Z]$ and $F, G \in S_6$ be such that $\langle F, G \rangle^\perp = I$. Since $(I_{\leq 4}) = (x^2, xy)$, we may assume that $F, G \in \langle Y^6, Y^5Z, Y^4Z^2, Y^3Z^3, Y^2Z^4, YZ^5, Z^6, XZ^5 \rangle$. So, let

$$\begin{aligned} F &= a_1 Y^6 + a_2 Y^5 Z + a_3 Y^4 Z^2 + a_4 Y^3 Z^3 + a_5 Y^2 Z^4 + a_6 Y Z^5 + a_7 Z^6 + a X Z^5, \\ G &= b_1 Y^6 + b_2 Y^5 Z + b_3 Y^4 Z^2 + b_4 Y^3 Z^3 + b_5 Y^2 Z^4 + b_6 Y Z^5 + b_7 Z^6. \end{aligned}$$

Since $\mathbf{H}(A, 1) = 3$ we cannot have both F and G in $k[Y, Z]_6$ and so we may assume that $a = 1$.

Now consider all the contractions of F and G. They may be viewed as vectors in k^7 as follows:

$$\begin{aligned}
D_X(F) &= Z^5 \\
&\leftrightarrow (0,0,0,0,0,1,0), \\
D_Y(F) &= a_1 Y^5 + a_2 Y^4 Z + a_3 Y^3 Z^2 + a_4 Y^2 Z^3 + a_5 Y Z^4 + a_6 Z^5 \\
&\leftrightarrow (a_1, a_2, a_3, a_4, a_5, a_6, 0), \\
D_Z(F) &= a_2 Y^5 + a_3 Y^4 Z + a_4 Y^3 Z^2 + a_5 Y^2 Z^3 + a_6 Y Z^4 + a_7 Z^5 + X Z^4 \\
&\leftrightarrow (a_2, a_3, a_4, a_5, a_6, a_7, 1), \\
D_X(G) &= 0 \\
&\leftrightarrow (0,0,0,0,0,0,0), \\
D_Y(G) &= b_1 Y^5 + b_2 Y^4 Z + b_3 Y^3 Z^2 + b_4 Y^2 Z^3 + b_5 Y Z^4 + b_6 Z^5 \\
&\leftrightarrow (b_1, b_2, b_3, b_4, b_5, b_6, 0), \\
D_Z(G) &= b_2 Y^5 + b_3 Y^4 Z + b_4 Y^3 Z^2 + b_5 Y^2 Z^3 + b_6 Y Z^4 + b_7 Z^5 \\
&\leftrightarrow (b_2, b_3, b_4, b_5, b_6, b_7, 0).
\end{aligned}$$

Hence

$$A = \begin{pmatrix} 0 & 0 & 0 & 0 & 0 & 1 & 0 \\ a_1 & a_2 & a_3 & a_4 & a_5 & a_6 & 0 \\ a_2 & a_3 & a_4 & a_5 & a_6 & a_7 & 1 \\ b_1 & b_2 & b_3 & b_4 & b_5 & b_6 & 0 \\ b_2 & b_3 & b_4 & b_5 & b_6 & b_7 & 0 \end{pmatrix}$$

has rank 4 since $H(R/I, 5) = 4$, and hence

$$A_1 = \begin{pmatrix} a_1 & a_2 & a_3 & a_4 & a_5 \\ b_1 & b_2 & b_3 & b_4 & b_5 \\ b_2 & b_3 & b_4 & b_5 & b_6 \end{pmatrix}$$

has rank 2.

We now consider the double contractions of F and G, where we identify a polynomial

$$\alpha_1 Y^4 + \alpha_2 Y^3 Z + \alpha_3 Y^2 Z^2 + \alpha_4 Y Z^3 + \alpha_5 Z^4 + \beta X Z^3$$

as a vector $(\alpha_1, \alpha_2, \alpha_3, \alpha_4, \alpha_5, \beta) \in k^6$. Then we have

$$
\begin{aligned}
D_{Z,X}(F) &= Z^4 \\
&\leftrightarrow (0,0,0,0,1,0), \\
D_{Y,Y}(F) &= a_1 Y^4 + a_2 Y^3 Z + a_3 Y^2 Z^2 + a_4 Y Z^3 + a_5 Z^4 \\
&\leftrightarrow (a_1, a_2, a_3, a_4, a_5, 0), \\
D_{Z,Y}(F) &= a_2 Y^4 + a_3 Y^3 Z + a_4 Y^2 Z^2 + a_5 Y Z^3 + a_6 Z^4 \\
&\leftrightarrow (a_2, a_3, a_4, a_5, a_6, 0), \\
D_{Z,Z}(F) &= a_3 Y^4 + a_4 Y^3 Z + a_5 Y^2 Z^2 + a_6 Y Z^3 + a_7 Z^4 + X Z^3 \\
&\leftrightarrow (a_3, a_4, a_5, a_6, a_7, 1), \\
D_{Y,Y}(G) &= b_1 Y^4 + b_2 Y^3 Z + b_3 Y^2 Z^2 + b_4 Y Z^3 + b_5 Z^4 \\
&\leftrightarrow (b_1, b_2, b_3, b_4, b_5, 0), \\
D_{Z,Y}(G) &= b_2 Y^4 + b_3 Y^3 Z + b_4 Y^2 Z^2 + b_5 Y Z^3 + b_6 Z^4 \\
&\leftrightarrow (b_2, b_3, b_5, b_4, b_6, 0), \\
D_{Z,Z}(G) &= b_3 Y^4 + b_4 Y^3 Z + b_5 Y^2 Z^2 + b_6 Y Z^3 + b_7 Z^4 \\
&\leftrightarrow (b_3, b_4, b_5, b_6, b_7, 0).
\end{aligned}
$$

Hence we have

$$
B = \begin{pmatrix}
0 & 0 & 0 & 0 & 1 & 0 \\
a_1 & a_2 & a_3 & a_4 & a_5 & 0 \\
a_2 & a_3 & a_4 & a_5 & a_6 & 0 \\
a_3 & a_4 & a_5 & a_6 & a_7 & 1 \\
b_1 & b_2 & b_3 & b_4 & b_5 & 0 \\
b_2 & b_3 & b_4 & b_5 & b_6 & 0 \\
b_3 & b_4 & b_5 & b_6 & b_7 & 0
\end{pmatrix}
$$

has rank 6 since $\mathbf{H}(R/I, 4) = 6$, and so

$$
B_1 = \begin{pmatrix}
a_1 & a_2 & a_3 & a_4 \\
a_2 & a_3 & a_4 & a_5 \\
b_1 & b_2 & b_3 & b_4 \\
b_2 & b_3 & b_4 & b_5 \\
b_3 & b_4 & b_5 & b_6
\end{pmatrix}
$$

has rank 4. Since the rank of A_1 is 2, one of three row vectors of A_1 has to be a linear combination of the other two. Checking the various possibilities for A_1, we find that the matrix B_1 can have rank at most 3, a contradiction.

Hence, the O-sequence $\mathbf{H} = (1, 3, 4, 5, 6, 4, 2)$ is not level.

Using the same ideas as above, we can show that any O-sequence of the form

$$h = (1, 3, 4, 5, \ldots, 6, 4, 2)$$

is not level.

5.2. Level Quotients of the Co-ordinate Rings of Points

Notice that Lemma 2.5 gives us a method of finding level quotients of any Artinian graded algebra. Unfortunately, Lemma 2.5 doesn't give much information about the Hilbert function of the level algebra it constructs.

In spite of that, M. Boij [**8**] (and later [**14**]) have explored a construction method similar to that of Lemma 2.5. We will now report on their construction method and then show that, in the case they consider, something can be said about the Hilbert function of the resulting level algebra.

The idea of Boij was to consider graded algebras which are the coordinate rings of sets of distinct points in $\mathbb{P}^n$ and to form level ideals in such rings. We look at that situation now.

Let $\mathbb{X} = \{P_1, \ldots, P_N\}$ be a collection of N distinct points in $\mathbb{P}^n$ and let $I_\mathbb{X}$ be the ideal generated by all the forms which vanish on all the points of $\mathbb{X}$. Then $I_\mathbb{X} = \wp_1 \cap \cdots \cap \wp_N$ where $\wp_i$ (the ideal of forms vanishing at P_i) is $\wp_i = (L_{i1}, \ldots, L_{in})$ (where the $L_{ij}, j = 1, \ldots, n$ are n linearly independent linear forms). The homogeneous coordinate ring of $\mathbb{X}$ will be denoted $A_\mathbb{X} = P/I_\mathbb{X}$, where $P = k[x_0, \ldots, x_n]$.

Now, the natural map from $A_\mathbb{X}$ to its integral closure (in the total ring of fractions of A) is given by taking the natural projection maps:

$$\pi_i : A_\mathbb{X} \to A_\mathbb{X}/\overline{\wp_i} = P/(L_{i1}, \ldots, L_{in}) \simeq k[T_i]$$

and putting them all together to get

$$\varphi = (\pi_1, \ldots, \pi_N), \ \varphi : A_\mathbb{X} \to \bigoplus_{i=1}^N (A_\mathbb{X}/\overline{\wp_i}) = \bigoplus_{i=1}^N k[T_i] := \mathcal{S}$$

Notice that φ is an injective, graded homomorphism (of degree 0). If we denote by $\sigma(\mathbb{X}) := \min\{ t \mid \Delta\mathbf{H}_\mathbb{X}(t) = 0 \}$, then

$$\varphi_d : (A_\mathbb{X})_d \to \mathcal{S}_d \text{ is an isomorphism for all } d \geq \sigma(\mathbb{X}) - 1.$$

If we choose a set of projective coordinates for each of the points $P_1, \ldots, P_N$, then, for each j,

$$\varphi_j(G) = \left(G(P_1)T_1^j, \ldots, G(P_N)T_N^j\right).$$

Now, with no loss of generality, we can always assume that $P_i = [1 : a_{i1} : \cdots : a_{in}]$, i.e., that all the P_i are in the affine piece of $\mathbb{P}^n$ which is the complement of the hyperplane $x_0 = 0$. With that observation we see that $\overline{x_0}$ is not a zero divisor in $A_\mathbb{X}$.

It is easy to find forms F_i of degree $\sigma(\mathbb{X}) - 1$, $i = 1, \ldots, N$ such that $F_i(P_j) = \delta_{ij}$. Then $\overline{F_1}, \ldots, \overline{F_N}$ are a basis for $(A_\mathbb{X})_{\sigma(\mathbb{X})-1}$ and so their images are a basis for $\mathcal{S}_{\sigma(\mathbb{X})-1}$. In fact their images are

$$\varphi_{\sigma(\mathbb{X})-1}(\overline{F_i}) = \left(0, \ldots, 0, T_i^{\sigma(\mathbb{X})-1}, 0, \ldots, 0\right).$$

It is also easy to see that $\{\overline{x_0}^\ell \overline{F_i}\}_{i=1}^N$ are a basis for $(A_\mathbb{X})_{\sigma(\mathbb{X})+\ell-1}$.

Now, let H be a subspace of $(A_\mathbb{X})_c$, $c \geq \sigma(\mathbb{X}) - 1$, of codimension ℓ. Then H has a basis of $N - \ell$ vectors, $\overline{G_1}, \ldots, \overline{G_{N-\ell}}$. We may consider these vectors as lying in $\mathcal{S}_c$ and write them as column vectors $v_1, \ldots, v_{N-\ell}$, where

$$v_1 = \begin{bmatrix} G_1(P_1) \\ \vdots \\ G_1(P_N) \end{bmatrix}, \ldots, v_{N-\ell} = \begin{bmatrix} G_{N-\ell}(P_1) \\ \vdots \\ G_{N-\ell}(P_N) \end{bmatrix}.$$

Let $M = [\lambda_{ij}]$ be an $\ell \times N$ matrix of rank ℓ for which

$$Mv_1 = \cdots = Mv_{N-\ell} = 0.$$

I.e., M is a matrix whose rows are a basis for the orthogonal complement of the subspace of $\mathcal{S}_c$ given by $\langle v_1, \ldots, v_{N-\ell} \rangle$.

In spite of the fact that M is not determined by H we will refer to M as *a matrix determined by H*. Note that the rowspace of M is determined by H.

It would be very useful to understand the kinds of level algebras that arise from the level ideals of subspaces $H \subset (A_{\mathbb{X}})_c$, where $A_{\mathbb{X}}$ is the coordinate ring of a set of points $\mathbb{X}$ in $\mathbb{P}^n$. A particular case to consider occurs when H describes those forms which vanish on some subset $\mathbb{Y}$ of $\mathbb{X}$. It would be very interesting if the geometry of $\mathbb{Y}$ in $\mathbb{X}$ could be used to describe the Hilbert function of such quotients of $A_{\mathbb{X}}$.

Fortunately, when c is large with respect to $\sigma(\mathbb{X})$ there is something we can say. We first recall a definition.

DEFINITION 5.10 ([**38**]). An O-sequence $(1, h_1, \ldots, h_s)$, with $h_s \neq 0$, is called *flawless* if

i) $h_i \leq h_{s-i}$ for any i, $0 \leq i \leq [s/2]$, and
ii) $h_1 \leq \cdots \leq h_{[s/2]}$.

Flawless O-sequences are not necessarily unimodal; consider $(1, 3, 5, 6, 5, 5, 6)$ which is flawless but not unimodal.

We now prove a proposition which gives some information about level k-algebras constructed using level ideals in coordinate rings of set of points in $\mathbb{P}^n$.

PROPOSITION 5.11. *Let $\mathbb{X}$ be a set of N points in $\mathbb{P}^n$ (as above) and choose $c \geq 2\sigma(\mathbb{X}) - 2$. Let $H \subset (A_{\mathbb{X}})_c$ have codimension ℓ and let $M = (\lambda_{ij})$ be an $\ell \times N$ matrix determined by H. Suppose that $\lambda_{1j} \neq 0$ for $j = 1, \ldots, N$. Let $A = A_{\mathbb{X}} / \langle H : A_{\mathbb{X}} \rangle$ be the level algebra quotient of $A_{\mathbb{X}}$ defined by H. Then, the Hilbert function of A is flawless.*

PROOF. Consider the codimension one subspace, H', of $(A_{\mathbb{X}})_c$ which is orthogonal to the *first* row of M. Since $H' \supset H$ we have that $\langle H : A_{\mathbb{X}} \rangle \subset \langle H' : A_{\mathbb{X}} \rangle$. Set $A' = A_{\mathbb{X}} / \langle H' : A_{\mathbb{X}} \rangle$. Then A' is Gorenstein with socle degree c. From [[**8**], Prop. 2.4] we can actually describe the Hilbert function of A'. It is

$$\mathbf{H}_{A'}(d) = \begin{cases} \mathbf{H}_{\mathbb{X}}(d), & \text{if } 0 \leq d \leq [c/2], \\ \mathbf{H}_{\mathbb{X}}(c-d), & \text{if } [c/2] \leq d \leq c. \end{cases}$$

Since $c \geq 2\sigma(\mathbb{X}) - 2$ we have $[c/2] \geq \sigma(\mathbb{X}) - 1$, i.e., the h-vector of A' is:

$$\left(1, \mathbf{H}_{\mathbb{X}}(1), \ldots, \mathbf{H}_{\mathbb{X}}(\sigma(\mathbb{X})-2), N, \ldots, N, \mathbf{H}_{\mathbb{X}}(\sigma(\mathbb{X})-2), \ldots, \mathbf{H}_{\mathbb{X}}(1), 1\right).$$

Since $\langle H : A_{\mathbb{X}} \rangle \subset \langle H' : A_{\mathbb{X}} \rangle$ we have:

$$\mathbf{H}_{A'}(d) \leq \mathbf{H}_A(d) \leq \mathbf{H}_{\mathbb{X}}(d) \quad \text{for every } d.$$

Since A' and $A_{\mathbb{X}}$ have the same Hilbert function for $d \leq c - (\sigma(\mathbb{X}) - 1)$ we get

(5.3) $\qquad \mathbf{H}_{A'}(d) = \mathbf{H}_A(d) = \mathbf{H}_{\mathbb{X}}(d) \quad \text{for all } d \leq c - (\sigma(\mathbb{X}) - 1)$

and consequently

$$1 = \mathbf{H}_A(0) \leq \mathbf{H}_A(1) \leq \cdots \leq \mathbf{H}_A([c/2])$$

since this is true for the reduced ring $A_{\mathbb{X}}$. Moreover, for $d \leq [c/2]$ we have

$$\mathbf{H}_A(d) = h_{A'}(d) = \mathbf{H}_{A'}(c-d) \leq \mathbf{H}_A(c-d).$$

This finishes the proof. $\square$

REMARK 5.12. *i*) It will be an easy consequence of Proposition 5.15 below that the ring A constructed in Proposition 5.11 also satisfies the WLP (and hence has unimodal h-vector).

ii) The set

$$\{(\lambda_{ij}) \in M_{\ell,N}(k) \mid \text{the rank of the matrix } (\lambda_{ij}) \text{ is equal to } \ell \text{ and } \lambda_{1j} \neq 0 \text{ for all } j\}$$

is a Zariski open subset of $k^{\ell N}$ considered as an affine space (where we identify $M_{\ell,N}(k)$ with $k^{\ell N}$). Thus, taking into account *i*), we see that the Hilbert functions of "most" Artinian level quotients (with fixed socle degree and type) of the coordinate ring of a finite set of points in $\mathbb{P}^n$ are flawless and unimodal, if $c \geq 2\sigma(\mathbb{X}) - 2$.

We are motivated to ask the following question about the "end" of the h-vector of a level algebra (with the WLP).

QUESTION 5.13. (1) Is the h-vector of a level algebra flawless?
(2) Is the h-vector of a level algebra of type two flawless?
(3) Is the h-vector of a level algebra with the WLP flawless?
(4) Is the h-vector of a level algebra, with the WLP, of type two flawless?

REMARK 5.14. Let $h = (1, n, h_2, \ldots, h_s)$ be the h-vector of a level algebra $A = \oplus_{i=0}^{s} A_i$ with the WLP, where $s \geq 2$. We will consider the SI-sequence $a = (1, n, a_2, \ldots, a_s)$ (see page 41 for the definition of an SI-sequence), where

$$a_i = \begin{cases} h_i & \text{for } i = 0, \ldots, [s/2], \\ h_{s-i} & \text{for } i = [s/2] + 1, \ldots, s. \end{cases}$$

Assume that h is flawless. Then we have $a_i \leq h_i$ for all $0 \leq i \leq s$. So, we would say that $a = (1, n, a_2, \ldots, a_s)$ is the maximal SI-sequence which is contained in $h = (1, n, h_2, \ldots, h_s)$. Consider the following natural question: With the same notation above, let $h_s = 2$; does there exist $f \in A_s$ such that the h-vector of the Gorenstein algebra $A/<f : A>$ coincides with the maximal SI-sequence a of h?

We were very surprised to realize that a beautiful example of Iarrobino [45] provides a counterexample. Indeed, he provides an explicit pair of polynomials F and G of degree 4 with the following properties. First, the Hilbert function of the level algebra, A, coming from the corresponding inverse system, is $(1, 3, 6, 6, 2)$ (note that this is flawless and even compressed!). Second, he shows that no Gorenstein quotient of A also having socle degree four is compressed Gorenstein. That is, not Gorenstein quotient has Hilbert function $(1, 3, 6, 3, 1)$. We have verified on the computer that Iarrobino's explicit example has the WLP, so it is indeed a counterexample.

5.3. Constructing Level Algebras with the WLP

There are some very simple ways to construct Artinian algebras with the WLP.

PROPOSITION 5.15. *Let $\mathbb{X}$ be a finite set of points in $\mathbb{P}^n$ and let A be an Artinian quotient of the coordinate ring of $\mathbb{X}$. Assume that $\mathbf{H}(A, i) = \mathbf{H}(\mathbb{X}, i)$ for all $0 \leq i \leq \sigma(\mathbb{X}) - 1$. Then A has the WLP. In particular, if $h = (1, h_1, \ldots, h_{s-1}, h_s)$ is an O-sequence such that $h_i = \mathbf{H}(\mathbb{X}, i)$ for all $i \leq s - 1$, where $\mathbb{X}$ is a set of h_{s-1} points, and $h_s \leq h_{s-1}$, then h is a level sequence consistent with WLP.*

5.3. CONSTRUCTING LEVEL ALGEBRAS WITH THE WLP

PROOF. Note that $\sigma(\mathbb{X}) - 1$ is the first degree for which $\mathbf{H}(\mathbb{X}, i) = |\mathbb{X}|$, the multiplicity of $\mathbb{X}$. Let $B = \oplus_{i \geq 0} B_i$ be the coordinate ring of $\mathbb{X}$, let $g \in B_1$ be a non zero divisor and let $c = \sigma(\mathbb{X}) - 1$. Consider the following commutative diagram:

$$\begin{array}{ccccccccc} B_0 & \xrightarrow{g} & B_1 & \xrightarrow{g} & \cdots & \xrightarrow{g} & B_c & \xrightarrow{g} & B_{c+1} & \xrightarrow{g} & \cdots \\ \downarrow \varphi & & \downarrow \varphi & & & & \downarrow \varphi & & \downarrow \varphi & & \\ A_0 & \xrightarrow{\overline{g}} & A_1 & \xrightarrow{\overline{g}} & \cdots & \xrightarrow{\overline{g}} & A_c & \xrightarrow{\overline{g}} & A_{c+1} & \xrightarrow{\overline{g}} & \cdots \end{array}$$

where φ is the canonical surjection from B to A.

Noting that $\mathbf{H}(A, i) = \mathbf{H}(\mathbb{X}, i)$ for all $0 \leq i \leq \sigma(\mathbb{X}) - 1$, it follows that $B_i \cong A_i$ for all $0 \leq i \leq c = \sigma(\mathbb{X}) - 1$. Hence, since $B_i \xrightarrow{g} B_{i+1}$ is injective for all i, we have that $A_i \xrightarrow{\overline{g}} A_{i+1}$ is injective for all $0 \leq i \leq c - 1 = \sigma(\mathbb{X}) - 2$. Also, since φ is surjective in all degrees and $B_i \xrightarrow{g} B_{i+1}$ is an isomorphism for all $i \geq c$, we obtain from the commutativity that $A_i \xrightarrow{\overline{g}} A_{i+1}$ is surjective in all degrees $i \geq c$, and hence we have WLP for A.

The final assertion comes from the observation that with the stated assumptions, $s - 1 \geq \sigma(\mathbb{X}) - 1$ so we can take $c = s - 1$ in the diagram above and take $A_{c+1} = A_s$ to be any quotient of $(R/I_{\mathbb{X}})_{c+1}$ of dimension h_s, (and $A_i = 0$ for all $i \geq c + 2$) and use essentially the same argument. $\square$

PROPOSITION 5.16. *Let $h = (1, h_1, h_2, \ldots, h_{s-2}, h_{s-1}, 1)$ be a symmetric O-sequence (i.e., $h_i = h_{s-i}$ for all i). Suppose also that*

$$(1, h_1 - 1, h_2 - h_1, \ldots, h_{[s/2]} - h_{[s/2]-1})$$

is again an O-sequence.

If v is any integer, $0 < v \leq s$, then $(1, h_1, \ldots, h_v)$ is the h-vector of a level set of points in $\mathbb{P}^{h_1}$ whose general Artinian reduction has the WLP.

PROOF. Recall that vectors like h are referred to in the literature as *SI-sequences* in honour of Stanley and Iarrobino (see [**31**]). It was shown in [[**55**], Theorem 1.1] that there is always a Gorenstein set of points $\mathbb{X}$ in $\mathbb{P}^{h_1}$ whose h-vector is h as above and whose general Artinian reduction has the WLP. By [[**25**], Lemma 2.3, c)] there is a subset $\mathbb{Y} \subset \mathbb{X}$ whose h-vector is $(1, h_1, \ldots, h_v)$. Let $B = \bigoplus_{i \geq 0} B_i$ and $A = \bigoplus_{i \geq 0} A_i$ be the homogeneous coordinate rings of $\mathbb{X}$ and $\mathbb{Y}$ respectively.

Since $\mathbb{X}$ and $\mathbb{Y}$ have the same Hilbert function up to degree v we have that the natural surjection $B \to A$ is an isomorphism $B_i \to A_i$ for all $i = 0, 1, \ldots, v$.

If we pass to the general Artinian reduction of both rings, $\overline{B}$ and $\overline{A}$ (by the "same" linear form) we obtain the diagram

$$\begin{array}{ccc} \overline{B_i} & \longrightarrow & \overline{B_{i+1}} \\ \downarrow & & \downarrow \\ \overline{A_i} & \longrightarrow & \overline{A_{i+1}} \end{array}$$

where the vertical maps correspond to the previously mentioned isomorphisms (now for $i + 1 \leq v$) and the horizontal maps correspond to multiplication by a general linear form. Thus, since $\overline{B}$ is a level algebra with the WLP the same is also true for $\overline{A}$. $\square$

COROLLARY 5.17. *Let $\mathbf{H} = (1, h_1, \ldots, h_s)$ be an O-sequence for which $(1, h_1 - 1, \ldots, h_s - h_{s-1})$ is again an O-sequence. Then $\mathbf{H}$ is the h-vector of an Artinian level algebra with the WLP.*

PROOF. We merely form the symmetric vector
$$\overline{h} = (1, h_1, \ldots, h_{s-1}, h_s, h_{s-1}, \ldots, h_1, 1)$$
and apply the proposition above. □

There is yet another way to construct Artinian level algebras with the WLP which we have found very useful.

PROPOSITION 5.18. *Let A be a level Artinian algebra with the WLP and let $L \in A_1$ be a Lefschetz element of A. Define*
$$t = \min\{i \mid \mathbf{H}_A(i) \geq \mathbf{H}_A(i+1)\} \quad and \quad u = \min\{i \mid \mathbf{H}_A(i) > \mathbf{H}_A(i+1)\}.$$
Let d be an integer, $0 \leq d \leq u - t$. Then $B = A/(0 : L^d)$ is again a level algebra with the WLP and the image of L in B is a Lefschetz element of B. Moreover, the Hilbert function of B is given by:
$$\mathbf{H}_B(i) = \begin{cases} \mathbf{H}_A(i), & \text{for } i = 0, \ldots, u - d, \\ \mathbf{H}_A(i+d), & \text{for } i = u - d + 1, \ldots, s - d. \end{cases}$$
Thus, the socle degree of B is $s - d$.

PROOF. Note that $u - t = 0$ is certainly possible (in which case the proposition gives nothing new) and the number $u - t + 1$ counts the number of times that the h-vector of A (necessarily unimodal) attains its largest value. The h-vector of B differs from that of A in that d of those largest values have been deleted from the h-vector of A.

Notice also that the ith graded piece of $(0 : L^d)$ is nothing more than the kernel of the linear map
$$L^d : A_i \longrightarrow A_{d+i}.$$
Since A has the WLP and we've chosen d such that $0 \leq d \leq u - t$, it follows that

(5.4) $\qquad L^d : A_i \longrightarrow A_{d+i}$ is $\begin{cases} \text{injective} & \text{for } i = 0, 1, \ldots, u - d, \\ \text{surjective} & \text{for } i = u - d + 1, \ldots. \end{cases}$

Since $B_i \simeq A_i/\text{Ker}[L^d : A_i \to A_{i+d}]$, we get that

(5.5) $\qquad B \simeq A_0 \oplus A_1 \oplus \cdots \oplus A_{u-d} \oplus A_{u+1} \oplus \cdots \oplus A_s.$

Thus, the socle degree of B is $s - d$ and we get the equality stated above on the Hilbert function of B.

We now show that B is level, i.e., $\text{soc}(B) = B_{s-d}$. Let $\overline{a} \in \text{soc}(B)_i$, $i < s - d$, where we can assume that $a \in A_i$ and $aA_1 \subset (0 : L^d)$. Then $L^d(aA_1) = 0$, i.e., $(aL^d)A_1 = 0$. Hence $aL^d \in \text{soc}(A)$. But, $\deg(aL^d) < (s-d)+d = s$ and hence, since A is level, $aL^d = 0$. But then $a \in (0 : L^d)$, i.e., $\overline{a} = 0$ in B and so $\text{soc}(B) = B_{s-d}$.

Finally we show that B has the WLP by showing that $\overline{L} : B_i \to B_{i+1}$ is either injective or surjective. But, from the identification (5.5) on B we have that the multiplication $\overline{L} : B \to B$ can be described as

$$A_0 \xrightarrow{L} A_1 \xrightarrow{L} \cdots \xrightarrow{L} A_{u-d} \xrightarrow{L^{d+1}} A_{u+1} \xrightarrow{L} A_{u+2} \xrightarrow{L} \cdots \xrightarrow{L} A_s .$$

The only part requiring comment is the map $A_{u-d} \stackrel{L^{d+1}}{\to} A_{u+1}$. But

$$A_{u-d} \stackrel{L^d}{\to} A_u \stackrel{L}{\to} A_{u+1}$$

is the composition of two surjective maps and hence is surjective. $\square$

We have shown in Proposition 5.18 that the flat part of a level sequence with the WLP can be shortened. In the following proposition, we will show that the flat part of special level sequences can be extended limitlessly.

PROPOSITION 5.19. (a) *Let $\mathbb{X}$ be a finite set of points in $\mathbb{P}^n$ and let A be an Artinian level quotient of the coordinate ring of $\mathbb{X}$ with socle degree s. Assume that $\mathbf{H}(A,i) = \mathbf{H}(\mathbb{X},i)$ for all $0 \leq i \leq \sigma(\mathbb{X}) - 1$. Let d be a positive integer and define*

$$u = \min\{i \mid \mathbf{H}(A,i) > \mathbf{H}(A,i+1)\}$$

and

$$h_i = \begin{cases} \mathbf{H}(A,i) & \text{for } i = 0,\ldots,u, \\ \mathbf{H}(A,u) & \text{for } i = u+1,\ldots,u+d, \\ \mathbf{H}(A,i-d) & \text{for } i = u+d+1,\ldots,s+d. \end{cases}$$

Then $(1, h_1, \ldots, h_{s+d})$ is the h-vector of an Artinian level algebra with the WLP.

(b) *Let $\mathcal{C} \subset \mathbb{P}^n$ be an arithmetically Cohen-Macaulay curve and let $\mathbb{X} \subset \mathcal{C}$ be a level set of points such that $\mathbf{H}(\mathbb{X},i) = \mathbf{H}(\mathcal{C},i)$ for all $0 \leq i \leq \sigma(\mathcal{C}) - 1$. Let $(1, h'_1, h'_2, \ldots, h'_s)$ be the h-vector of $\mathbb{X}$ and let*

$$u = \min\{i \mid h'_i > h'_{i+1}\}.$$

Let d be a positive integer and define

$$h_i = \begin{cases} h'_i & \text{for } i = 0,\ldots,u, \\ h'_u & \text{for } i = u+1,\ldots,u+d, \\ h'_{i-d} & \text{for } i = u+d+1,\ldots,s+d. \end{cases}$$

Then $(1, h_1, h_2, \ldots, h_{s+d})$ is the h-vector of a level set of points on $\mathcal{C}$ whose Artinian reduction has the WLP.

PROOF. Our proof for part (b) below could be modified to also prove (a), but we record here a different proof for (a) to illustrate a different technique.

Let $B = \oplus_{i \geq 0} B_i$ be the coordinate ring of $\mathbb{X}$, let $H \subset B_s$ be a subspace of B_s such that $A = B/\langle H : B \rangle$ and let $g \in B_1$ be a non zero divisor of degree one.

We first show that

(5.6) $$g^d \langle H : B \rangle_i = \langle g^d H : B \rangle_{i+d}$$

for all $i \geq \sigma(\mathbb{X}) - 1$. Let $b \in \langle H : B \rangle_i$. Then, since $bB_{s-i} \subset H$, we have $g^d b B_{s+d-(i+d)} = g^d b B_{s-i} \subset g^d H$. Hence $g^d b \in \langle g^d H : B \rangle_{i+d}$. Conversely, let $b' \in \langle g^d H : B \rangle_{i+d}$. Then, since g^d is a non zero divisor, we have that the linear map $g^d : B_i \to B_{i+d}$ is bijective for all $i \geq \sigma(\mathbb{X}) - 1$. Hence, noting that $b' = g^d b$ for some $b \in B_i$, we see that $g^d b B_{s-i} = b' B_{s+d-(i+d)} \subset g^d H$. Again, noting that g^d is a non zero divisor, we have $b B_{s-i} \subset H$, i.e., $b \in \langle H : B \rangle_i$. Therefore, $b' \in g^d \langle H : B \rangle_i$.

From the assumption, $\mathbf{H}(A,i) = \mathbf{H}(\mathbb{X},i)$ for all $0 \leq i \leq \sigma(\mathbb{X}) - 1$, it follows that $\langle H : B \rangle_i = (0)$ for all $0 \leq i \leq u$, because $\sigma(\mathbb{X}) - 1 \leq u$ and $\mathbf{H}(A,i) = |\mathbb{X}|$ for all $\sigma(\mathbb{X}) - 1 \leq i \leq u$. Hence, from (5.6), we have

$$\langle g^d H : B \rangle_j = (0) \tag{5.7}$$

for all $0 \leq j \leq u + d$. Furthermore, since g^d is a non zero divisor, we see that $\dim \langle H : B \rangle_i = \dim g^d \langle H : B \rangle_i$ for all $i \geq 0$, Hence, from (5.6), we have

$$\dim \langle H : B \rangle_i = \dim \langle g^d H : B \rangle_{i+d} \tag{5.8}$$

for all $i \geq \sigma(\mathbb{X}) - 1$. Let $A' = B/\langle g^d H : B \rangle$. Then, noting that $\mathbf{H}(A,i) = \mathbf{H}(\mathbb{X},i)$ for all $0 \leq i \leq u$, it follows from (5.7) that

$$\mathbf{H}(A', i) = \begin{cases} \mathbf{H}(A, i) & \text{for } i = 0, \ldots, u, \\ \mathbf{H}(A, u) & \text{for } i = u+1, \ldots, u+d. \end{cases}$$

That is, $\mathbf{H}(A', i) = h_i$ for all $0 \leq i \leq u + d$. Furthermore, it follows from (5.8) that $\mathbf{H}(A', i) = \mathbf{H}(A, i-d) = h_i$ for all $u + d + 1 \leq i \leq s + d$. From Proposition 5.13, we have that A' has the WLP. This finishes the proof of (a).

Now we prove (b). Let F be a sufficiently general form of degree d. Let $\mathbb{Y}$ be the hypersurface section of $\mathcal{C}$ cut out by F. We will show that the union, $\mathbb{Z}$, of $\mathbb{X}$ and $\mathbb{Y}$ is the desired level set of points. This is in fact a so-called "basic double G-link" (see [**48**], Lemma 4.8 and its proof). The relevant facts that we will use are that the saturated ideal of $\mathbb{Z} = \mathbb{X} \cup \mathbb{Y}$ is $I_\mathbb{Z} = I_\mathcal{C} + F \cdot I_\mathbb{X}$, and there is a short exact sequence

$$0 \to I_\mathcal{C}(-d) \to I_\mathcal{C} \oplus I_\mathbb{X}(-d) \to I_\mathbb{Z} \to 0, \tag{5.9}$$

where the first non-trivial map sends $A \mapsto (FA, A)$ and the second non-trivial map sends $(A, B) \mapsto A - FB$.

Consider minimal free resolutions

$$\begin{aligned} \mathbb{G}_\bullet &: 0 \to \mathbb{G}_{n-1} \to \cdots \to \mathbb{G}_1 \to I_\mathcal{C} \to 0 \\ &\text{and} \\ \mathbb{F}_\bullet &: 0 \to \mathbb{F}_n \to \cdots \to \mathbb{F}_1 \to I_\mathbb{X} \to 0. \end{aligned} \tag{5.10}$$

Note that $I_\mathcal{C}$ is generated in degree $\leq \sigma(\mathcal{C})$. Then the condition that $\mathbf{H}(\mathbb{X}, i) = \mathbf{H}(\mathcal{C}, i)$ for all $0 \leq i \leq \sigma(\mathcal{C}) - 1$ means that in degrees $\leq \sigma(\mathcal{C}) - 1$, the generators of $I_\mathcal{C}$ and the generators of $I_\mathbb{X}$ coincide. Furthermore, in degree $\sigma(\mathcal{C})$, every minimal generator of $I_\mathcal{C}$ is also a minimal generator of $I_\mathbb{X}$ (but there may be other generators of $I_\mathbb{X}$). But this implies that in addition, every minimal k-th syzygy of $I_\mathcal{C}$ is also a minimal k-th syzygy of $I_\mathbb{X}$.

Now, placing the minimal free resolutions (5.10) vertically over the short exact sequence (5.9) in the obvious way, the mapping cone provides a free resolution (not minimal) for $I_\mathbb{Z}$. But the above observations give us the additional fact that *every* free module in $\mathbb{G}_\bullet(-d)$ splits off with the corresponding summands of $\mathbb{F}_\bullet(-d)$, so that we in fact have the precise minimal free resolution of $I_\mathbb{Z}$, and in particular it is level.

Now we verify the claim about the Hilbert function. We note the following facts.

(i) $\mathbf{H}(\mathbb{Y}, i) = \mathbf{H}(\mathcal{C}, i) - \mathbf{H}(\mathcal{C}, i - d)$. In particular, if $i \geq \sigma(\mathcal{C}) - 1 + d$ then $\Delta \mathbf{H}(\mathbb{Y}, i) = 0$.

(ii) Because $\Delta \mathbf{H}(\mathcal{C}, i) = \deg \mathcal{C}$ for all $i \geq \sigma(\mathcal{C}) - 1$, we have that in fact $\mathbf{H}(\mathbb{X}, i) = \mathbf{H}(\mathcal{C}, i)$ for all $i \leq u$, and also $\sigma(\mathcal{C}) - 1 \leq u$.

(iii) Consequently, $h'_i = \Delta \mathbf{H}(\mathcal{C}, i)$ for all $i \leq u$. Also, $h'_i = h'_u$ for all $\sigma(\mathcal{C}) - 1 \leq i \leq u$.

Now, from the sequence (5.9) it is easy to see that

$$\begin{aligned} \mathbf{H}(\mathbb{Z}, i) &= \mathbf{H}(\mathcal{C}, i) - \mathbf{H}(\mathcal{C}, i-d) + \mathbf{H}(\mathbb{X}, i-d) \\ &= \mathbf{H}(\mathbb{Y}, i) + \mathbf{H}(\mathbb{X}, i-d) \end{aligned}$$

For $i \leq u+d$, we have $\mathbf{H}(\mathcal{C}, i-d) = \mathbf{H}(\mathbb{X}, i-d)$ (from (ii)). Hence $\mathbf{H}(\mathbb{Z}, i) = \mathbf{H}(\mathcal{C}, i)$ for all $i \leq u+d$ (using (i)). Now we compute:

- For $0 \leq i \leq u$,

$$\begin{aligned} \Delta \mathbf{H}(\mathbb{Z}, i) &= \Delta \mathbf{H}(\mathcal{C}, i) \\ &= \Delta \mathbf{H}(\mathbb{X}, i) \quad \text{(from (ii))} \\ &= h'_i. \end{aligned}$$

- For $u+1 \leq i \leq u+d$,

$$\begin{aligned} \Delta \mathbf{H}(\mathbb{Z}, i) &= \Delta \mathbf{H}(\mathcal{C}, i) \\ &= \deg \mathcal{C} \quad \text{(since } u+1 \geq \sigma(\mathcal{C})\text{)} \\ &= \Delta \mathbf{H}(\mathbb{X}, u) \\ &= h'_u. \end{aligned}$$

- For $u+d+1 \leq i \leq s+d$,

$$\begin{aligned} \Delta \mathbf{H}(\mathbb{Z}, i) &= \Delta \mathbf{H}(\mathbb{Y}, i) + \Delta \mathbf{H}(\mathbb{X}, i-d) \\ &= \Delta \mathbf{H}(\mathbb{X}, i-d) \quad \text{(by (ii) and (i))} \\ &= h'_{i-d}. \end{aligned}$$

We will see below, in Proposition 6.17, that the Artinian reduction of $\mathbb{Z}$ has the WLP. $\square$

The following question asks whether we really need points in the first part of Proposition 5.19.

QUESTION 5.20. Let $h' = (1, n, h'_2, \ldots, h'_s)$ be the h-vector of a level algebra with the WLP, and let

$$u = \min\{i \mid h'_i > h'_{i+1}\}.$$

Let d be a positive integer and define

$$h_i = \begin{cases} h'_i & \text{for } i = 0, \ldots, u, \\ h'_u & \text{for } i = u+1, \ldots, u+d, \\ h'_{i-d} & \text{for } i = u+d+1, \ldots, s+d. \end{cases}$$

Then, is $(1, n, h_2, \ldots, h_s)$ the h-vector of a level algebra with the WLP?

REMARK 5.21. Using Proposition 5.19 above, we obtain special examples of level sequences with a long flat part. As we will see in the appendices, every level sequence in Table 6.2 of the appendices is produced by the linked-sum method satisfying the condition "$\mathbf{H}(A, i) = \mathbf{H}(\mathbb{X}, i)$ for all $0 \leq i \leq \sigma(\mathbb{X}) - 1$." Hence, for example, we can get the following level sequences with 208] and 326] in Table 6.2:

$$1, 3, 6, 7, 8, 8, 8, \ldots, 8, 8, 6, 2$$
$$1, 3, 6, 10, 10, 10, \ldots, 10, 10, 9, 6, 2, \quad \text{etc..}$$

Another simple way to get level algebras with the WLP is the following.

PROPOSITION 5.22. *Suppose that $\binom{n-1+t}{t} = nr$ for some integer r. Then, any compressed level algebra, A, with h-vector*

$$h = \left(1, n, \binom{n-1+2}{2}, \ldots, \binom{n-1+t}{t} = h_t, r\right)$$

satisfies the WLP.

PROOF. We can write $A = R/I$, with $R = k[x_1, \ldots, x_n]$, and our assumption on h implies that $A_i = R_i$ for $0 \le i \le t$. Thus, if L is *any* linear form in A_1, $L : A_i \to A_{i+1}$ is injective for $0 \le i \le t - 1$. It will be enough to show that $L : A_t \to A_{t+1}$ is surjective.

If it were not, then $L : A_{t+1}^* \to A_t^*$ would have a kernel. Thus, the canonical surjection (A is level), $A_1 \otimes A_{t+1}^* \to A_t^*$ would also have a kernel. But, by assumption $(\dim_k A_1)(\dim_k A_{t+1}^*) = \dim_k A_t^*$. This is the desired contradiction. Thus, *every* $L \in A_1$ is a Lefschetz element in A. □

REMARK 5.23. We can use this to show that the compressed algebras with h-vector $(1, 3, 6, 10, 15, 5)$ and $(1, 3, 6, 10, 15, 21, 7) \cdots$ all have the WLP.

One of the strongest results we have for finding level algebras which satisfy the WLP comes from re-examining a special case of Proposition 5.5. It is the following.

PROPOSITION 5.24. *Let $(1, h_1, h_2, \ldots, h_s)$ be the h-vector of a level algebra $A = R/I$ where $R = k[x_1, \ldots, x_n]$. Let $j = \max\left\{i \,\middle|\, h_i = \binom{i+n-1}{i}\right\}$. Then,*

  i) *there is an ideal $J \subset I$ so that $B = R/J$ is a level algebra with h- vector $(1, h_1, \ldots, h_j, h_{j+1} + 1, \ldots, h_s + 1)$;*
  ii) *moreover, if A satisfies the WLP then J can be chosen so that B also satisfies the WLP.*

PROOF. Note that i) is a very special case of Proposition 5.5. We will include a proof anyway since the notation will be useful for ii), which is new.

Let $S = k[y_1, \ldots, y_n]$ and let $I^{-1} \subset S$ be the inverse system of I. Then

$$\dim_k(I^{-1})_{j+1} = h_{j+1} < \dim_k S_{j+1}.$$

Let $V = (I^{-1})_{j+1}$ and choose linear forms $L_1, \ldots, L_{\binom{j+n}{j+1}}$ in S_1 so that $L_1^{j+1}, \ldots, L_{\binom{j+n}{j+1}}^{j+1}$ form a basis for S_{j+1} where, with no loss of generality, we can assume that $L_1^{j+1} \notin V$ and that the coefficient of y_1 in L_1 is $c_1 \neq 0$.

It follows that $L_1^{j+2} \notin (I^{-1})_{j+2}$, for if it were then we would have that

$$\frac{\partial}{\partial y_1}\left(L_1^{j+2}\right) = c_1(j+2)L_1^{j+1} \in V.$$

But, $c_1 \neq 0$ would then imply that $L_1^{j+1} \in V$, which is a contradiction.

Continuing in this way we see that $L_1^{j+(s-j)} \notin (I^{-1})_s = W$. Since $A = R/I$ was level, we know that $I^{-1} = \langle W \rangle$. So, if we consider the inverse system $\langle W, L_1^s \rangle = J^{-1}$ we get that $J \subset I$ and $B = R/J$ is again level and has the required h-vector.

ii) Let $J_1 = \operatorname{ann}_R(L_1^s)$. Then the h-vector of R/J_1 is $(1, \ldots, 1)$, where the last 1 occurs in degree s. It follows that J_1 is actually a complete intersection ideal of R, generated by $n - 1$ linearly independent forms of degree 1 and 1 form of degree $s + 1$. These $n - 1$ linear forms generate the prime ideal $\wp$ of a point in $\mathbb{P}^{n-1}$.

Now $J = I \cap J_1$ but, given our observation about the generators of J_1 we have, in degrees $\leq s$, that $J = I \cap \wp$. Consider the short exact sequence

(5.11) $$0 \to R/(I \cap \wp) \to R/I \oplus R/\wp \to R/(I + \wp) \to 0 .$$

It follows from $i)$ and (5.11) that

$$\dim_k (R/(I+\wp))_t = \begin{cases} 1, & \text{for } t \leq j, \\ 0, & \text{for } j < t \leq s, \end{cases}$$

and hence that, for $j < t \leq s$ we have a functorial isomorphism

(5.12) $$(R/J)_t \simeq (R/I)_t \oplus (R/\wp)_t.$$

Consider the maps

(5.13) $$\mu_t : (R/J)_t \to (R/J)_{t+1}, \quad \nu_t : (R/I)_t \to (R/I)_{t+1}$$

induced by multiplication by a general linear form, L. We assume that ν_t has maximal rank, and we want to show that μ_t has maximal rank. We will use the following diagram.

(5.14)
$$\begin{array}{ccc} (R/J)_t & \xrightarrow{\mu_t} & (R/J)_{t+1} \\ \downarrow r_t & & \downarrow r_{t+1} \\ (R/I)_t & \xrightarrow{\nu_t} & (R/I)_{t+1} \\ \downarrow & & \downarrow \\ 0 & & 0 \end{array}$$

We now consider four different possibilities.

Case 1. $t + 1 \leq j$. Then $(R/J)_{t+1} = (R/I)_{t+1} = R_{t+1}$ and the assertion is trivial.

Case 2. $t \geq s$. This case is trivial.

Case 3. $j < t < s$. In this case all four entries of (5.14) get increased by 1.

Suppose first that ν_t is injective. If μ_t is not injective, let $F \in \text{Ker}\mu_t$. There are two possibilities.
- If $r_t(F) \neq 0$ in $(R/I)_t$ then the injectivity of ν_t gives $\nu_t(r_t(F)) \neq 0$ while $r_{t+1}(\mu_t(F)) = 0$, contradicting the commutativity of the diagram.
- If $r_t(F) = 0$ in $(R/I)_t$ then from (5.11) and (5.12) we have that F corresponds to an ordered pair (α, β) with $\alpha = 0$ in $(R/I)_t$ but $\beta \neq 0$ in $(R/\wp)_t$. Since clearly multiplication by L is an isomorphism between consecutive components of $R/\wp$, we get that $\mu_t(F) \neq 0$. Contradiction.

  Now suppose that ν_t is surjective.

  Let F be a non-zero element of $(R/J)_{t+1}$. Thanks to (5.12), we can identify F with an ordered pair (α, β) where $\alpha \in (R/I)_{t+1}$ and $\beta \in (R/\wp)_{t+1}$. But multiplication by L is surjective on R/I and $R/\wp$ from degree t to degree $t + 1$, so $F \in \text{Im}(\mu_t)$.

Case 4. $t = j$. In this case, we have $(R/I)_t = R_t$ and $(R/J)_t = R_t$ but $(R/I)_{t+1} \neq R_{t+1}$. In particular, r_t is an isomorphism. Again most of the argument focuses on the diagram (5.14).

If $\nu_t : (R/I)_t \to (R/I)_{t+1}$ is an injection, suppose that $0 \neq F \in \text{Ker}(\mu_t) \subset (R/J)_t = R_t$. Since r_t is an isomorphism, $r_t(F) \neq 0$ in $(R/I)_t = R_t$. But then the injectivity of ν_t gives $\nu_t(r_t(F)) \neq 0$ while $r_{t+1}(\mu_t(F)) = r_{t+1}(0) = 0$, contradicting the commutativity of (5.14). Therefore μ_t is injective.

Now suppose that $\nu_t : (R/I)_t \to (R/I)_{t+1}$ is surjective.

- If ν_t is an isomorphism then $\nu_t \circ r_t$ is an isomorphism, so also $r_{t+1} \circ \mu_t$ is an isomorphism. Hence μ_t is injective.
- Assume that ν_t is surjective but not an isomorphism. Let $F \in (R/J)_{t+1}$. We distinguish between two subcases:

  * If $F \in \text{Ker}(r_{t+1})$, let G be any non-zero element of $\text{Ker}\nu_t$, which exists by the assumption that ν_t is not an isomorphism. Since r_t is an isomorphism, we may view $G \in (R/J)_t = R_t$. Then the element $\mu_t(G)$ is also in $\text{Ker}(r_{t+1})$. But $\text{Ker}(r_{t+1}) = (I/J)_{t+1}$ is 1-dimensional.

    We now **claim** that, without loss of generality, we may assume that $\mu_t(G)$ is non-zero in $\text{Ker}(r_{t+1})$. It is clear that every such G is automatically in $\text{Ker}(r_{t+1})$, so what we are really claiming is that there exists a $G \in \text{Ker}(\nu_t) \subset (R/I)_t = R_t = (R/J)_t$ such that $\mu_t(G) \neq 0$ in $(R/J)_{t+1}$. Let us use the information that we have acquired to rewrite (5.14). Recall that L is a general linear form, and μ_t and ν_t are the maps defined by multiplication by L. Recall also that here $t = j$.

$$\begin{array}{ccc}
R_t & \xrightarrow{\mu_t} & (R/I)_{t+1} \oplus (R/\wp)_{t+1} \\
{\scriptstyle r_t} \downarrow & & \downarrow {\scriptstyle r_{t+1}} \\
0 \longrightarrow (I:L)_t \longrightarrow R_t & \xrightarrow{\nu_t} & (R/I)_{t+1} \longrightarrow 0
\end{array}$$

    Note that $\text{Ker}(\nu_t) = (I:L)$ is independent of the choice of $\wp$. Without loss of generality we may choose $\wp$ so that any given $G \in (I:L)$ does not vanish at the point corresponding to $\wp$. Similarly we may assume that L is not in $\wp$. Therefore LG is not zero in $R/\wp$, so $G \notin \text{Ker}(\mu_t)$.

    The above argument proves the claim. Therefore, for some $\lambda \in K$ (the ground field), $F - \mu_t(\lambda G) = 0$ and $F \in \text{Im}(\mu_t)$.

  * Now let F be any element of $(R/J)_{t+1}$. Let $G \in (R/I)_t$ be a non-zero element such that $\nu_t(G) = r_{t+1}(F)$. Again we may view G as an element of $(R/J)_t = R_t$. Now $F - \mu_t(G) \in \text{Ker}(r_{t+1})$. By the previous subcase, then, $F - \mu_t(G) \in \text{Im}(\mu_t)$. But then also $F \in \text{Im}(\mu_t)$.

Therefore μ_t is surjective, and Case 4 is complete.

These four cases complete the proof that R/J has the WLP. $\square$

5.4. The Linked-Sum Method

Another very useful construction method for Artinian level algebras having embedding dimension $n + 1$ requires the construction of level sets of points in $\mathbb{P}^n$. This is usually referred to as the ***linked-sum*** method and is based on the following observations: if $\mathbb{Z} = \mathbb{X} \bigcup \mathbb{Y}$ is a reduced set of points in $\mathbb{P}^n$ ($\mathbb{X}$ and $\mathbb{Y}$ are disjoint subsets of $\mathbb{Z}$) then: if $\mathbb{Z}$ is a level set of points and $R = k[x_0, \ldots, x_n]$ then $A = R/(I_\mathbb{X} + I_\mathbb{Y})$ is an Artinian level algebra.

Moreover, the Hilbert function of A is given by

$$\mathbf{H}(A, t) = \mathbf{H}(R/I_\mathbb{X}, t) + \mathbf{H}(R/I_\mathbb{Y}, t) - \mathbf{H}(R/I_\mathbb{Z}, t) .$$

(For details and some applications of this construction see [**23**]).

Obviously, the usefulness of this construction depends on first being able to find level sets of points and then to find such which have subsets with a large variety of Hilbert functions. I.e., these have to be rather special sets of level points.

In [**23**] we indicated a very specific construction to find level sets of points in $\mathbb{P}^2$. We recall that construction here since we want to show that one can say even more than was mentioned in [**23**] about the resulting level algebras. We recall some definitions.

DEFINITION 5.25 ([**33**]). We let $R = k[x, y, z]$ denote the coordinate ring of $\mathbb{P}^2$.
1) A finite set $\mathbb{X}$ of points in $\mathbb{P}^2$ is called a *basic configuration* of type (d, e) if there exist distinct elements b_j, c_j in k such that

$$I_{\mathbb{X}} = \left(\prod_{j=1}^{d}(x - b_j z), \prod_{j=1}^{e}(y - c_j z) \right).$$

We write $\mathbb{X} := \mathbb{B}(d, e)$ and think of the points of $\mathbb{X}$ as being the 'lattice points' of a rectangle with e rows and d columns (note the order).
2) A finite set $\mathbb{X}$ of points in $\mathbb{P}^2$ is called a *pure configuration* if there exist finite basic configurations $\mathbb{B}(d_1, e_1), \ldots, \mathbb{B}(d_m, e_m)$ where $e_1 > \cdots > e_m$, which satisfy the following three conditions.

  i) $\mathbb{B}(d_i, e_i) \cap \mathbb{B}(d_j, e_j) = \varnothing$ if $i \neq j$,
  ii) $\mathbb{X} = \mathbb{B}(d_1, e_1) \bigcup \cdots \bigcup \mathbb{B}(d_m, e_m)$,
  iii) $\varphi(\mathbb{B}(d_i, e_i)) \supset \varphi(\mathbb{B}(d_{i+1}, e_{i+1}))$ for all $1 \leq i \leq m-1$, where $\varphi : \mathbb{P}^2 \setminus \{[1 : 0 : 0]\} \to \mathbb{P}^1$ is the map defined by sending the point $[x : y : z]$ to the point $[y : z]$.

In this case, we denote $\mathbb{X} = \bigcup_{i=1}^{m} \mathbb{B}(d_i, e_i)$.

REMARK 5.26. Since basic configurations are (very special) complete intersections, their Hilbert functions are trivial to calculate.

In [[**33**], Lemma 4.4] the Hilbert function of a pure configuration was calculated in terms of the Hilbert functions of its basic components. More precisely, if

$$\mathbb{Z} = \bigcup_{i=1}^{m} \mathbb{B}(d_i, e_i)$$

and the numbers $v_0, v_1, \ldots, v_m$ are defined by $v_0 = 0$ and $v_j = d_1 + \cdots + d_j$, then

(5.15) $$\mathbf{H}(\mathbb{Z}, i) = \sum_{j=1}^{m} \mathbf{H}(\mathbb{B}(d_j, e_j), i - v_{j-1}).$$

Moreover, in [[**23**], Corollary 3.10] we showed

(5.16) $\mathbb{Z}$ as above is a level set of points $\iff e_i - e_{i+1} = d_{i+1}, 1 \leq i \leq m-1$.

From this remark we see that at least one of the ingredients necessary to use the "linked-sum" approach has been collected, i.e., we have an easy construction for lots of sets of points in $\mathbb{P}^2$ which are level.

Now, to get level Artinian algebras from these level sets of points (perhaps which have even more special properties) we need to decide how to split up these pure configurations.

The following lemma describes what happens when we split the level pure configuration by, basically, splitting one of the basic configurations into two smaller basic configurations and partitioning the pure configuration with a "fence" between these two pieces of the broken up basic configuration. More precisely.

LEMMA 5.27. *Let* $\mathbb{Z} = \bigcup_{i=1}^{m} \mathbb{B}(d_i, e_i)$ *be a pure configuration in* $\mathbb{P}^2$, *and consider a partition of* $\mathbb{Z}$ *as follows.* $\mathbb{Z} = \mathbb{Z}' \bigcup \mathbb{Z}''$, *where*

$$\mathbb{Z}' = \left\{ \bigcup_{i=1}^{\ell-1} \mathbb{B}(d_i, e_i) \right\} \bigcup \mathbb{B}(d'_\ell, e_\ell)$$

$$\text{and} \quad \mathbb{Z}'' = \mathbb{B}(d''_\ell, e_\ell) \bigcup \left\{ \bigcup_{i=\ell+1}^{m} \mathbb{B}(d_i, e_i) \right\}$$

such that $d'_\ell + d''_\ell = d_\ell$ *and* $\mathbb{B}(d'_\ell, e_\ell) \bigcup \mathbb{B}(d''_\ell, e_\ell) = \mathbb{B}(d_\ell, e_\ell)$. *(We consider* $\mathbb{B}(d''_\ell, e_\ell) = \varphi$ *if* $d''_\ell = 0$.*) Then*

$$\mathbf{H}(\mathbb{Z}, i) = \mathbf{H}(\mathbb{Z}', i) + \mathbf{H}(\mathbb{Z}'', i - (d_1 + \cdots + d_{\ell-1} + d'_\ell))$$

for all $i \geq 0$.

PROOF. First note that since $\mathbb{B}(d_\ell, e_\ell)$ is a complete intersection, we can use the Cayley Bacharach Theorem (see [**15**]) to say that

$$\Delta\mathbf{H}(\mathbb{B}(d_\ell, e_\ell), i) = \Delta\mathbf{H}(\mathbb{B}(d'_\ell, e_\ell), i) + \Delta\mathbf{H}(\mathbb{B}(d''_\ell, e_\ell), d_\ell + e_\ell - 2 - i).$$

Since $d_\ell + e_\ell - 2 - i = d'_\ell + d''_\ell + e_\ell - 2 - i$ we have

$$\Delta\mathbf{H}(\mathbb{B}(d''_\ell, e_\ell), d_\ell + e_\ell - 2 - i) = \Delta\mathbf{H}(\mathbb{B}(d''_\ell, e_\ell), d'_\ell + d''_\ell + e_\ell - 2 - i).$$

Now, using the symmetry of the difference function of the Hilbert function of a complete intersection, we get that

$$\Delta\mathbf{H}(\mathbb{B}(d''_\ell, e_\ell), d'_\ell + d''_\ell + e_\ell - 2 - i) = \Delta\mathbf{H}(\mathbb{B}(d''_\ell, e_\ell), d''_\ell + e_\ell - 2 - (d'_\ell + d''_\ell + e_\ell - 2 - i))$$

and this last is nothing more than

$$\Delta\mathbf{H}(\mathbb{B}(d''_\ell, e_\ell), i - d'_\ell).$$

Hence we get the identity,

(5.17) $$\mathbf{H}(\mathbb{B}(d_\ell, e_\ell), i) = \mathbf{H}(\mathbb{B}(d'_\ell, e_\ell), i) + \mathbf{H}(\mathbb{B}(d''_\ell, e_\ell), i - d'_\ell).$$

Now, since $\mathbb{Z}'$ and $\mathbb{Z}''$ are both pure configurations, consider the integers (see Remark 5.26) $v'_0, v'_1, \ldots, v'_\ell$ associated to $\mathbb{Z}'$ and $v''_0, v''_1, \ldots, v''_{m-\ell+1}$ associated to $\mathbb{Z}''$. Notice that $v'_0 = v_0 = 0, v'_1 = v_1, \ldots, v'_{\ell-1} = v_{\ell-1}$ and $v'_\ell = v_{\ell-1} + d'_\ell$.

Also, $v''_0 = 0$, $v''_1 = d''_\ell$, $v''_2 = d''_\ell + d_{\ell+1}, \ldots, v''_{m-\ell+1} = d''_\ell + d_{\ell+1} + \cdots + d_m$. I.e.,

$$v''_i = d''_\ell + d_{\ell+1} + \cdots + d_{\ell+i-1}$$

which we can rewrite as

$$v''_i = v_{\ell+i-1} - v_{\ell-1} - d'_\ell.$$

From (5.15), we have

$$\mathbf{H}(\mathbb{Z}', i) = \sum_{j=1}^{\ell-1} \mathbf{H}(\mathbb{B}(d_i, e_i), i - v'_{j-1}) + \mathbf{H}(\mathbb{B}(d'_\ell, e_\ell), i - v'_{\ell-1})$$

and

$$\mathbf{H}(\mathbb{Z}'', i) = \mathbf{H}(\mathbb{B}(d''_\ell, e_\ell), i) + \sum_{j=1}^{m-\ell} \mathbf{H}(\mathbb{B}(d_{\ell+j}, e_{\ell+j}), i - v''_j).$$

Now consider

$$\mathbf{H}(\mathbb{Z}'', i - (d_1 + \cdots + d_{\ell-1} + d'_\ell)) = \mathbf{H}(\mathbb{Z}'', i - (v_{\ell-1} + d'_\ell))$$

5.4. THE LINKED-SUM METHOD

which, in view of what we wrote above, can be rewritten as

$$\mathbf{H}(\mathbb{B}(d_\ell'', e_\ell), i - (v_{\ell-1} + d_\ell')) + \sum_{j=1}^{m-\ell} \mathbf{H}(\mathbb{B}(d_{\ell+j}, e_{\ell+j}), i - (v_{\ell-1} + d_\ell') - v_j'') .$$

Taking advantage of our way of rewriting v_j'', this last is

$$\mathbf{H}(\mathbb{B}(d_\ell'', e_\ell), i - v_\ell') + \sum_{j=1}^{m-\ell} \mathbf{H}(\mathbb{B}(d_{\ell+j}, e_{\ell+j}), i - v_{\ell+j-1}) .$$

Putting this together with (5.17) gives us the conclusion of the lemma. □

Now recall that a k-configuration in $\mathbb{P}^2$ of type $\mathcal{T} = (r_1, r_2, \ldots, r_u)$, $r_1 < r_2 < \cdots < r_u$ is a collection of $r = \sum_{i=1}^u r_i$ distinct points of $\mathbb{P}^2$ situated on distinct lines $\mathbb{L}_1, \ldots, \mathbb{L}_u$ of $\mathbb{P}^2$ in such a way that exactly r_i of the points lie on the line $\mathbb{L}_i$ and no point chosen on line $\mathbb{L}_i$ lies on the line $\mathbb{L}_j$ for $j < i$. We have the following proposition.

PROPOSITION 5.28. *Let $\mathbb{Z} = \bigcup_{i=1}^m \mathbb{B}(d_i, e_i)$ be a pure configuration in $\mathbb{P}^2$ satisfying $e_i - e_{i+1} = d_{i+1}$ for all $i = 1, 2, \ldots, m-1$. Let $\mathcal{T} = (r_1, \ldots, r_u)$ be a 2-type vector such that $r_u = e_1$, $r_u - r_{u-1} \geq e_2$ and $u \leq d_1$, and let $\mathbb{X}$ be a k-configuration of type $\mathcal{T}$ which is contained in $\mathbb{B}(d_1, e_1)$.*

Set $\mathbb{Y} := \mathbb{Z} - \mathbb{X}$, $\mathbb{Y}' := \mathbb{B}(d_1, e_1) - \mathbb{X}$, $\mathbb{Y}'' := \bigcup_{i=2}^m \mathbb{B}(d_i, e_i)$, and $h_i = \mathbf{H}(R/(I_\mathbb{X} + I_\mathbb{Y}), i)$ for all $i = 0, 1, \ldots$. Then we have the following.

1) *$R/(I_\mathbb{X} + I_\mathbb{Y})$ is an Artinian level graded k-algebra of C-M type m;*
2) *the socle degrees of the rings $R/(I_\mathbb{X} + I_\mathbb{Y}))$ and $R/(I_\mathbb{X} + I_{\mathbb{Y}'})$ are both $d_1 + e_1 - 3$;*
3) *the Hilbert function of $R/(I_\mathbb{X} + I_\mathbb{Y})$ is*

$$h_i = \begin{cases} \mathbf{H}(R/(I_\mathbb{X} + I_{\mathbb{Y}'}), i), & \text{for } i = 0, 1, \ldots, d_1 - 2, \\ \mathbf{H}(R/(I_\mathbb{X} + I_{\mathbb{Y}'}), i) + \Delta \mathbf{H}(\mathbb{Y}'', i - (d_1 - 1)), & \text{for } i = d_1 - 1, \ldots, d_1 + e_1 - 3. \end{cases}$$

Furthermore assume that $e_1 \leq d_1 - 1$. Then

4)

$$h_i = \begin{cases} \mathbf{H}(\mathbb{X}, i), & \text{for } i = 0, 1, \ldots, d_1 - 2, \\ \mathbf{H}(\mathbb{X}, d_1 + e_1 - 3 - i) & \text{for } i = d_1 - 1, \ldots, \\ + \Delta \mathbf{H}(\mathbb{Y}'', i - (d_1 - 1)), & d_1 + e_1 - 3, \end{cases}$$

and

5) *$R/(I_\mathbb{X} + I_\mathbb{Y})$ has the weak Lefschetz property.*

PROOF. First we note from [[23], Corollaries 3.10 and 3.15] that $R/(I_\mathbb{X} + I_\mathbb{Y})$ is an Artinian level graded k-algebra of C-M type less than or equal to m and whose socle degree is equal to $d_1 + e_1 - 3$.

From the usual exact sequence

$$0 \to R/I_\mathbb{Z} \to R/I_\mathbb{X} \oplus R/I_\mathbb{Y} \to R/(I_\mathbb{X} + I_\mathbb{Y}) \to 0,$$

we have that

$$\mathbf{H}(R/(I_\mathbb{X} + I_\mathbb{Y}), i) = \mathbf{H}(\mathbb{X}, i) + \mathbf{H}(\mathbb{Y}, i) - \mathbf{H}(\mathbb{Z}, i)$$

for all $i \geq 0$. In the same way, we see that

$$\mathbf{H}(R/(I_\mathbb{X} + I_{\mathbb{Y}'}), i) = \mathbf{H}(\mathbb{X}, i) + \mathbf{H}(\mathbb{Y}', i) - \mathbf{H}(\mathbb{B}(d_1, e_1), i)$$

for all $i \geq 0$. Furthermore, it follows from Lemma 5.27 that
$$\mathbf{H}(\mathbb{Z},i) = \mathbf{H}(\mathbb{B}(d_1,e_1),i) + \mathbf{H}(\mathbb{Y}'',i-d_1).$$

Since $\mathbb{Y}$ is a pure configuration and $r_u - r_{u-1} \geq e_2$, we can use Lemma 5.27 to partition $\mathbb{Y}$ as $\mathbb{Y} = \mathbb{Y}' \bigcup \mathbb{Y}''$. I.e., noting that $r_u = e_1$, we have that
$$\mathbf{H}(\mathbb{Y},i) = \mathbf{H}(\mathbb{Y}',i) + \mathbf{H}(\mathbb{Y}'',i-(d_1-1)).$$

Hence, from the four equalities above, we get
$$\begin{aligned} h_i &= \mathbf{H}(\mathbb{X},i) + \{\mathbf{H}(\mathbb{Y}',i) + \mathbf{H}(\mathbb{Y}'',i-(d_1-1))\} \\ &\quad - \{\mathbf{H}(\mathbb{B}(d_1,e_1),i) + \mathbf{H}(\mathbb{Y}'',i-d_1)\} \\ &= \mathbf{H}(R/(I_{\mathbb{X}}+I_{\mathbb{Y}'}),i) + \Delta\mathbf{H}(\mathbb{Y}'',i-(d_1-1)). \end{aligned}$$

Now, applying [[31], Theorem 2.1 (3)] we get that the socle degree of $R/(I_{\mathbb{X}}+I_{\mathbb{Y}'})$ (i.e., $s(R/(I_{\mathbb{X}}+I_{\mathbb{Y}'}))$) is $d_1 + e_1 - 3$, and so $s(R/(I_{\mathbb{X}}+I_{\mathbb{Y}})) = s(R/(I_{\mathbb{X}}+I_{\mathbb{Y}'}))$. Furthermore, since $e_i - e_{i+1} = d_{i+1}$ for all $i = 1, 2, \ldots, m-1$, it follows from [[33], Lemma 4.4 (2)], that $\sigma(\mathbb{Y}'') = d_2 + e_2 - 1 = e_1 - 1$. Hence we see that $(d_1 - 1) + (\sigma(\mathbb{Y}'') - 1) = s(R/(I_{\mathbb{X}}+I_{\mathbb{Y}'}))$.

This gives the equalities of 3).

In particular, when $i = d_1 + e_1 - 3$, we have
$$\begin{aligned} \mathbf{H}(R/(I_{\mathbb{X}}+I_{\mathbb{Y}}),d_1+e_1-3) &= \mathbf{H}(R/(I_{\mathbb{X}}+I_{\mathbb{Y}'}),d_1+e_1-3) + \Delta\mathbf{H}(\mathbb{Y}'',e_1-2) \\ &= 1 + (m-1) \\ &= m. \end{aligned}$$

Therefore, from the first part of this proof it follows that the C-M type of $R/(I_{\mathbb{X}}+I_{\mathbb{Y}})$ is exactly equal to m. This completes the proof of 1), 2) and 3).

If we note that $\sigma(\mathbb{X}) = r_u = e_1$ and $\sigma(\mathbb{B}(d_1,e_1)) = d_1 + e_1 - 1$, then assertion 4) follows from [[31], Theorem 2.1 (4)].

As for 5), notice that $\sigma(\mathbb{X}) = e_1$ and $e_1 \leq d_1 - 1$ and thus $\sigma(\mathbb{X}) - 1 \leq d_1 - 2$. Hence, it follows from 4) that $\mathbf{H}(R/I_{\mathbb{X}} + I_{\mathbb{Y}}, i) = \mathbf{H}(\mathbb{X},i)$ for all i with $0 \leq i \leq \sigma(\mathbb{X}) - 1$. Thus, by Proposition 5.15, $R/I_{\mathbb{X}} + I_{\mathbb{Y}}$ has the WLP. □

REMARK 5.29. Let $\underline{h}$ be a Gorenstein O-sequence with $h_1 = 3$ and $\sigma(\underline{h}) = s+1$. Then, as was shown by Stanley, the "first half" of the O-sequence is a differentiable O-sequence. Let $\mathcal{T} = (r_1, r_2, \ldots, r_u)$ be the 2-type vector of this differentiable O-sequence associated to $\underline{h}$ (for more details see the Definition in [[31], page 318].)

Set $d_1 := s+3-r_u$ and $e_1 := r_u$. Let $\mathbb{Z} = \bigcup_{i=1}^{m} \mathbb{B}(d_i,e_i)$ be a pure configuration satisfying $e_i - e_{i+1} = d_{i+1}$ for all $i = 1, 2, \ldots, m-1$, and let $\mathbb{X}$ be a k-configuration of type $\mathcal{T}$ which is contained in $\mathbb{B}(d_1, e_1)$. (From [[31], Theorem 3.3], we have that there is always such a k-configuration in $\mathbb{B}(d_1, e_1)$.)

Set $\mathbb{Y} := \mathbb{Z} - \mathbb{X}$, $\mathbb{Y}' := \mathbb{B}(d_1,e_1) - \mathbb{X}$, and $\mathbb{Y}'' := \bigcup_{i=2}^{m} \mathbb{B}(d_i,e_i)$. Notice that from (5.16) above, $\mathbb{Y}''$ is a level set of points in $\mathbb{P}^2$. Let $\underline{a} = (a_0, a_1, \ldots, a_t)$ be the h-vector of $\mathbb{Y}''$. If $r_u - r_{u-1} \geq e_2$, then from [[31], Theorem 3.3], [[32], Lemma 3.2] and Proposition 5.28 above, we have the following.

1) $R/(I_{\mathbb{X}}+I_{\mathbb{Y}'})$ is an Artinian Gorenstein k-algebra with the weak Lefschetz property and the h-vector of $R/(I_{\mathbb{X}}+I_{\mathbb{Y}'})$ is equal to $\underline{h}$;
2) $R/(I_{\mathbb{X}}+I_{\mathbb{Y}})$ is an Artinian level k-algebra with CM type m with the weak Lefschetz property and whose h-vector is equal to
$$(h_0, h_1, \ldots, h_{s-t-1}, h_{s-t}+a_0, h_{s-t+1}+a_1, \ldots, h_s+a_t).$$

5.4. THE LINKED-SUM METHOD

The level sequences of codimension three obtained by this construction will be described as a *sum of an Artinian Gorenstein-sequence of codimension three and a level sequence of codimension two*.

We illustrate the discussion above with some examples.

EXAMPLE 5.30. Let $\underline{h} = (1, 3, 5, 6, 5, 3, 1)$. $\mathcal{T} = (2, 4)$ is the 2-type vector associated to the O-sequence 1 3 5 6 6 $\rightarrow$ (which is the differentiable O-sequence associated to $\underline{h}$).

Let $\mathbb{X}$ be the set of all •'s in the basic configuration $\mathbb{B}(5, 4)$ below and $\mathbb{Y}'$ the set of all $*$'s in that $\mathbb{B}(5, 4)$. Then $\mathbb{X}$ is a k-configuration of type $\mathcal{T} = (2, 4)$ and $\mathbb{Y}'$ is a pure configuration.

```
•  *  *  *  *
•  *  *  *  *
•  •  *  *  *
•  •  *  *  *
```

Now choose two positive integers d_2 and e_2 such that $d_2 + e_2 = 4$,

$$(d_2, e_2) = \{(1, 3), (2, 2), (3, 1)\},$$

and take a pure configuration $\mathbb{Z} := \mathbb{B}(5, 4) \bigcup \mathbb{B}(d_2, e_2)$ such that $\mathbb{Y} := \mathbb{Y}' \bigcup \mathbb{B}(d_2, e_2)$ is also a pure configuration.

To apply the results above we need $4 - 2 = 2 \geq e_2$ and so we have to eliminate the choice $(d_2, e_2) = (1, 3)$. For the other two choices we get, when $(d_2, e_2) = (2, 2)$.

```
•  *  *  *  *  *  *
•  *  *  *  *  *  *
•  •  *  *  *
•  •  *  *  *
```

Then the Hilbert functions of $\mathbb{X}$, $\mathbb{Y}$, $\mathbb{Z}$, and $R/(I_{\mathbb{X}} + I_{\mathbb{Y}})$ (using the notation of Remark 5.29) are

$$
\begin{array}{rcccccccccc}
\mathbf{H}(\mathbb{Z}, -) & : & 1 & 3 & 6 & 10 & 14 & 18 & 22 & 24 & \rightarrow \\
\mathbf{H}(\mathbb{X}, -) & : & 1 & 3 & 5 & 6 & 6 & 6 & 6 & 6 & \rightarrow \\
\mathbf{H}(\mathbb{Y}, -) & : & 1 & 3 & 6 & 10 & 14 & 17 & 18 & 18 & \rightarrow \\
\mathbf{H}(R/(I_{\mathbb{X}} + I_{\mathbb{Y}}), -) & : & 1 & 3 & 5 & 6 & 6 & 5 & 2 & 0 & \rightarrow.
\end{array}
$$

From Proposition 5.28 4) above, we get that $(1, 3, 5, 6, 6, 5, 2)$ is a level O-sequence and the h- vector of a level algebra with the WLP.

Let $(d_2, e_2) = (3, 1)$.

```
•  *  *  *  *  *  *
•  *  *  *  *
•  •  *  *  *
•  •  *  *  *
```

Then the Hilbert functions of $\mathbb{X}$, $\mathbb{Y}$, $\mathbb{Z}$, and $R/(I_{\mathbb{X}} + I_{\mathbb{Y}})$ (again with the notation of Remark 5.29) are

$$
\begin{array}{rcccccccccc}
\mathbf{H}(\mathbb{Z}, -) & : & 1 & 3 & 6 & 10 & 14 & 18 & 21 & 23 & \rightarrow \\
\mathbf{H}(\mathbb{X}, -) & : & 1 & 3 & 5 & 6 & 6 & 6 & 6 & 6 & \rightarrow \\
\mathbf{H}(\mathbb{Y}, -) & : & 1 & 3 & 6 & 10 & 14 & 16 & 17 & 17 & \rightarrow \\
\mathbf{H}(R/(I_{\mathbb{X}} + I_{\mathbb{Y}}), -) & : & 1 & 3 & 5 & 6 & 6 & 4 & 2 & 0 & \rightarrow.
\end{array}
$$

From Proposition 5.28, 4) above, we get that $(1, 3, 5, 6, 6, 4, 2)$ is a level O-sequence and the h-vector of a level algebra with the WLP.

QUESTION 5.31. Suppose that $(1, h_1, \ldots, h_s)$ is a Gorenstein sequence with $h_1 = 3$. Choose t so that $t \leq \max\{i \mid h_i < h_{i+1}\}$ and suppose that $(1, a_1, \ldots, a_t)$ is a level sequence with $a_1 = 2$. Suppose further that the sequence

$$(1, h_1, \ldots, h_{s-t-1}, h_{s-t} + 1, h_{s-t+1} + a_1, \ldots, h_s + a_t)$$

is unimodal (the sequence is obviously flawless). Is this a level sequence?

REMARK 5.32. Unfortunately, the linked sum construction cannot always be used to construct level Artinian algebras of type > 1 in codimension 3 from level sets of points in $\mathbb{P}^2$ (in sharp contrast to the case of Gorenstein algebras of codimension 3, see [[**31**], Theorem 3.3]).

An easy way to see this is to consider potential h-vectors describing level Artinian algebras of socle degree 5 and type 4. For example, let $h = (1, 3, -, -, \alpha, 4)$. By Inverse Systems we know that $\alpha \leq 12$ and, e.g. that there is a compressed level algebra with h-vector $(1, 3, 6, 10, 12, 4)$. However, that h-vector and indeed any h-vector of the form $(1, 3, -, -, \geq 11, 4)$ cannot be obtained from the linked-sum construction.

To see why, suppose we had a level set of points $\mathbb{Z} \subset \mathbb{P}^2$ and a partition $\mathbb{Z} = \mathbb{X} \bigcup \mathbb{Y}$ which gave the h-vector above by the linked-sum construction. Then, if we let $A = R/(I_\mathbb{X} + I_\mathbb{Y})$, we would have a display

$\mathbf{H}(\mathbb{Z}, -)$:	1	–	–	–	b	c
$\mathbf{H}(\mathbb{X}, -)$:	1	–	–	–	α_1	α_2
$\mathbf{H}(\mathbb{Y}, -)$:	1	–	–	–	β_1	β_2
$\mathbf{H}(A, -)$:	1	–	–	–	s	4

where $11 \leq s \leq 12$.

The construction method says that $b + s = \alpha_1 + \beta_1$ and $c + 4 = \alpha_2 + \beta_2$. Subtracting, we have

$$(c - b) + (4 - s) = (\alpha_2 - \alpha_1) + (\beta_2 - \beta_1).$$

But, since $\mathbb{X}$ and $\mathbb{Y}$ are point sets in $\mathbb{P}^2$ we have $(\alpha_2 - \alpha_1) \geq 0$ and $(\beta_2 - \beta_1) \geq 0$. Since $4 - s \leq -7$ we must have $c - b \geq 7$. But, since $\mathbb{Z}$ has a differentiable Hilbert function, the maximum value for $c - b$ is $6 = \dim_k(k[x, y]_5)$.

A very similar discussion, but this time applied to the compressed level h-vector $(1, 3, 6, 10, 15, 20, 12, 6, 2)$, shows that since $\dim_k(k[x, y]_6) = 7$ and $(12 - 20) = -8$, a compressed level algebra with this h-vector cannot be constructed by the linked sum method.

CHAPTER 6

Constructing Level Sets of Points

Another extremely interesting way to construct Artinian level k-algebras of codimension n is to construct level sets of points in $\mathbb{P}^n$. This is, by far, the strongest approach since by constructing level sets of points in $\mathbb{P}^n$ we obtain both Artinian level algebras of codimension n (taking the Artinian reduction of the coordinate ring of the set of points) and Artinian level algebras of codimension $n+1$ (using the "linked-sum" method).

We have had some success with this approach – enough so that we wonder if the following question might have a positive answer.

QUESTION 6.1. If $h = (1, n, h_2, \ldots, h_s)$, $h_s \neq 0$, is the h-vector of an Artinian level quotient of $R = k[x_1, \ldots, x_n]$ then is there a set of points in $\mathbb{P}^n$ whose coordinate ring has Artinian reduction with h-vector equal to h?

It is well known that Question 6.1 is true when $n = 2$. It was also shown in [28] to be true when $n = 3$ and $h_s = 1$. We will give an affirmative answer to Question 6.1 when $n = 3$ and $r \leq 4$ and also when $n = 3$, $s = 5$, 6 and the level algebra is of type 2. Those computations will all be put in the Appendix to this monograph.

In this monograph we will concentrate on constructing level sets of points in $\mathbb{P}^3$. It will be clear (and we will comment when this is so) that our constructions can be generalized to make examples in $\mathbb{P}^n$ for $n > 3$. But, since there are still so many open questions in codimension 3, we will restrict most of our attention there.

To produce level sets of points in $\mathbb{P}^3$, we have four main techniques.
1) Find them as the intersection of two suitable arithmetically Cohen-Macaulay curves with no common component whose union is a complete intersection and whose scheme-theoretic intersection is a reduced set of points. (This is the analogue, one dimension higher, of the linked-sum construction mentioned above.)
2) Find them by a liaison method, as described below.
3) Find them as a suitable union of smaller sets of points. We have one method, and we wonder if other similar methods exist.
4) Find them as a subset of a suitable curve in $\mathbb{P}^3$. Sometimes this can be used in conjunction with liaison.

6.1. Method 1: Intersection of Suitable Curves

A variant of this technique was used by the first and third authors to construct all possible sets of graded Betti numbers for arithmetically Gorenstein sets of points in $\mathbb{P}^3$ (see [28]). It was based on the observation that if $\mathbb{Z}$ is a complete intersection *stick figure* (i.e. a reduced union of lines where at most two meet in any point) in

$\mathbb{P}^3$, and if $\mathbb{X}$ and $\mathbb{Y}$ are both arithmetically Cohen-Macaulay curves linked by $\mathbb{Z}$, then $I_\mathbb{X} + I_\mathbb{Y}$ is the saturated ideal of a reduced set of points which is arithmetically Gorenstein, and which has computable Hilbert function and graded Betti numbers.

The crux of the problem in the more general case of level sets of points, is to mimic this construction and find *arithmetically Cohen-Macaulay* stick figures whose union, $\mathbb{Z}$ is level of type ≥ 2. In the Gorenstein situation, it sufficed to find one, and then the residual in $\mathbb{Z}$ was automatically arithmetically Cohen-Macaulay as well. In the current case this is no longer true. We will give one partial result which has proven useful in constructing many of our examples. First we need to recall a construction.

LEMMA 6.2 ([55]). *Let $\mathbb{V}_1 \subset \mathbb{V}_2 \subset \cdots \subset \mathbb{V}_t \subset \mathbb{P}^n$ be arithmetically Cohen-Macaulay schemes of the same dimension, each generically Gorenstein. Let $\mathbb{H}_1, \ldots, \mathbb{H}_t$ be hypersurfaces, defined by forms $F_1, \ldots, F_t$, such that for each i, $\mathbb{H}_i$ contains no component of $\mathbb{V}_j$ for any $j \leq i$. Let $\mathbb{W}_i$ be the arithmetically Cohen-Macaulay schemes defined by the corresponding hypersurface sections: $I_{\mathbb{W}_i} = I_{\mathbb{V}_i} + (F_i)$. Then*

i) *viewed as divisors on $\mathbb{V}_t$, the sum $\mathbb{Z}$ of the $\mathbb{W}_i$ (which is just the union, if the hypersurfaces are general enough) is in the same Gorenstein liaison class as $\mathbb{W}_1$.*

ii) *In particular, $\mathbb{Z}$ is arithmetically Cohen-Macaulay.*

iii) *As ideals we have*

$$I_\mathbb{Z} = I_{\mathbb{V}_t} + F_t \cdot I_{\mathbb{V}_{t-1}} + F_t F_{t-1} I_{\mathbb{V}_{t-2}} + \cdots + F_t F_{t-1} \cdots F_2 I_{\mathbb{V}_1} + (F_t F_{t-1} \cdots F_1).$$

iv) *Let $d_i = \deg F_i$. The Hilbert functions are related by the formula*

$$\begin{aligned} h_\mathbb{Z}(x) &= h_{\mathbb{W}_t}(x) + h_{\mathbb{W}_{t-1}}(x - d_t) + h_{\mathbb{W}_{t-2}}(x - d_t - d_{t-1}) + \cdots + \\ &\quad h_{\mathbb{W}_1}(x - d_t - d_{t-1} - \cdots - d_2). \end{aligned}$$

REMARK 6.3. 1) Notice that any way we partition the $\mathbb{W}_i$'s into two subsets whose union is $\mathbb{Z}$, each of the components is arithmetically Cohen-Macaulay. This is one of the more useful consequences of this lemma for us.

2) Parts *ii)*, *iii)* and *iv)* of this Lemma have been proved independently by Ragusa and Zappalà [[62], Lemma 1.5].

Now let $L_1, \ldots, L_d$ and $M_1, \ldots, M_e$ be families of linear forms in $R = k[x_0, \ldots, x_n]$, and assume that the ideal (A, B) in R is a complete intersection, where $A = \prod_{i=1}^d L_i$ and $B = \prod_{k=1}^e M_k$.

We use Lemma 6.2 in the following way.

COROLLARY 6.4. *Thinking of the L_i and M_k as hyperplanes which are pairwise linearly independent, the intersection of any L_i with M_k is a codimension two linear variety, $\mathbb{P}_{i,k}$.*

Consider a union $\mathbb{Y}$ of such codimension two linear varieties subject to the condition that if $\mathbb{P}_{i,k} \subset \mathbb{Y}$ then $\mathbb{P}_{j,\ell} \subset \mathbb{Y}$ for all (j, ℓ) satisfying $j \leq i$ and $\ell \leq k$.

Then $\mathbb{Y}$ is arithmetically Cohen-Macaulay.

If $n = 3$ and if any ideal of the form (L_i, L_j, M_k, M_l) is (x_0, x_1, x_2, x_3)-primary, then any subset of this complete intersection is a stick figure (independent of whether or not it is arithmetically Cohen-Macaulay).

6.1. METHOD 1: INTERSECTION OF SUITABLE CURVES

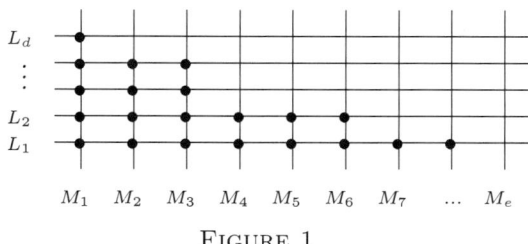

FIGURE 1

REMARK 6.5. If, for example, we consider Corollary 6.4 for the case $n = 3$, then the •'s in Figure 1 above correspond to lines of $\mathbb{P}^3$. One should not think of the hyperplanes corresponding to the L_i and L_j as being, in any way, parallel to each other. The diagram gives, at best, a rough idea of what is going on geometrically.

We summarize our use of Method 1 with the following result. Note, though, that it can be extended to higher projective space, and if desired it can be extended to produce level sets of points with type bigger than two.

COROLLARY 6.6. *Let $n = 3$ and let $\{\mathbb{P}_{i,k}\}$ ($1 \leq i \leq d, 1 \leq k \leq e$) be the complete intersection defined by (A, B) as above. Inside of $\{\mathbb{P}_{i,k}\}$, let $\mathbb{W}$ be a complete intersection of type (a, a) ($0 < a < d, 0 < a < e$). Let $\mathbb{Z}$ be the residual to $\mathbb{W}$ in the larger complete intersection $\{\mathbb{P}_{i,k}\}$. Then the Artinian reduction of $\mathbb{Z}$ is level of type 2 with socle degree $d + e - a - 2$. Furthermore, $\mathbb{Z}$ is a stick figure.*

Suppose that $\mathbb{Z} = \mathbb{X} \bigcup \mathbb{Y}$, where $\mathbb{X}$ and $\mathbb{Y}$ are unions of lines satisfying the "nested" condition of Corollary 6.4. Then $I_\mathbb{X} + I_\mathbb{Y}$ is the saturated ideal of a reduced union of points that is level of socle degree $d + e - a - 3$.

REMARK 6.7. We can get examples with socle of dimension > 2 if we remove more than one complete intersection from the original complete intersection. I.e., if the set $\mathbb{Z}$ in Corollary 6.6 looks like a pure configuration as described by (5.16) above.

EXAMPLE 6.8. Let $n = 3$ and consider a complete intersection $(5, 5)$ given by linear forms (see the diagram in Corollary 6.4) $L_1, \ldots, L_5$, $M_1, \ldots, M_5$.

We wish to apply Corollary 6.6, so let $\mathbb{W}$ be a complete intersection $(1, 1)$ and $\mathbb{Z}$ the residual to $\mathbb{W}$. Write $\mathbb{Z} = \mathbb{X} \bigcup \mathbb{Y}$, where $\mathbb{X}$ consists of lines represented by •'s and $\mathbb{Y}$ the lines represented by ∗'s in Figure 2.

$$\mathbb{Z} = \begin{Bmatrix} \bullet & \bullet & \bullet & \bullet & \\ \bullet & \ast & \ast & \ast & \ast \\ \ast & \ast & \ast & \ast & \ast \\ \ast & \ast & \ast & \ast & \ast \\ \ast & \ast & \ast & \ast & \ast \end{Bmatrix}$$

FIGURE 2

Notice that if we label the rows $\{1, 2, 3, 4, 5\}$ (from top to bottom) and the columns by $\{1, 2, 3, 4, 5\}$ (from left to right) we see that the $\mathbb{X}$'s satisfy the nested condition of Corollary 6.4 and so the lines they represent have ACM coordinate ring.

If, however, we label the rows $\{1, 2, 3, 4, 5\}$ (this time from bottom to top) and the columns $\{5, 4, 3, 2, 1\}$ (from left to right) we see that the $\mathbb{Y}$'s satisfy the nested condition of Corollary 6.4 and so they also represent lines whose homogeneous coordinate ring is ACM.

If we choose the lines generally enough the entire complete intersection will be a stick figure and hence the same will be true for $\mathbb{X}$ and $\mathbb{Y}$. That is enough to guarantee that $I_\mathbb{X} + I_\mathbb{Y}$ is the saturated ideal of a reduced level set of points in $\mathbb{P}^3$ of type 2 and whose Artinian reduction has socle degree $5 + 5 - 1 - 3 = 6$.

Calculating the Hilbert functions needed we get

$$\begin{array}{llcccccccc}
\Delta \mathbf{H}(R/I_\mathbb{Z}, -) & : & 1 & 3 & 6 & 10 & 15 & 19 & 22 & 24 & \to \\
\Delta \mathbf{H}(R/I_\mathbb{X}, -) & : & 1 & 3 & 4 & 5 & 5 & 5 & 5 & 5 & \to \\
\Delta \mathbf{H}(R/I_\mathbb{Y}, -) & : & 1 & 3 & 6 & 10 & 14 & 17 & 19 & 19 & \to \\
\Delta \mathbf{H}(R/I_\mathbb{X} + I_\mathbb{Y}, -) & : & 1 & 3 & 4 & 5 & 4 & 3 & 2 & 0 & \to.
\end{array}$$

So, the h-vector $(1, 3, 4, 5, 4, 3, 2)$ is a level sequence which comes from a level set of points in $\mathbb{P}^3$.

We will gather many other examples using this construction in the Appendix.

Notice also that the constructions used in Example 5.30 are both adaptable to Method 1.

6.2. Method 2: Liaison Tricks

Liaison has proven to be a powerful method of constructing interesting objects in projective space, and here we find a new example of its utility. We refer to [**53**] or [**56**] for the background needed for this discussion.

DEFINITION 6.9. A subscheme $\mathbb{V}$ of $\mathbb{P}^n$ is said to be *licci* if it is in the (complete intersection) **l**iaison **c**lass of a **c**omplete **i**ntersection. $\mathbb{V}$ is said to be *glicci* if it is in the **G**orenstein **l**iaison **c**lass of a **c**omplete **i**ntersection.

The *mapping cone construction* (see the above references) is a very useful way of obtaining a free resolution of a linked variety (in the case where the varieties are arithmetically Cohen-Macaulay), given a free resolution of the original variety. Although it usually does not give a minimal free resolution, often additional information can be used to split off redundant terms and obtain a minimal free resolution.

We illustrate this with an example.

EXAMPLE 6.10. Suppose we want to construct a level set of points, $\mathbb{Z}$, in $\mathbb{P}^3$ with h-vector $(1, 3, 6, 10, 12, 6, 2)$.

This is the h-vector of a compressed Artinian level algebra (see the comments after Proposition 5.4), i.e., in the parameter space of all Artinian level algebras of type 2 with socle degree 6, a Zariski open subset has this Hilbert function. We are, however, looking for **points** with this h-vector.

To produce a set of points in $\mathbb{P}^3$ with this h-vector we start with a set of 12 general points, $\mathbb{Z}_1$, in $\mathbb{P}^3$. This has h-vector $(1, 3, 6, 2)$ and, if $R = k[x_0, \ldots, x_3]$, minimal free resolution

(6.1) $$0 \to R(-6)^2 \to R(-4)^9 \to R(-3)^8 \to I_{\mathbb{Z}_1} \to 0.$$

6.2. METHOD 2: LIAISON TRICKS

Choose a general regular sequence J_1 of type (3,3,4) in $I_{\mathbb{Z}_1}$. J_1 has a minimal free resolution

(6.2) $$0 \to R(-10) \to \begin{bmatrix} R(-6) \\ \oplus \\ R(-7)^2 \end{bmatrix} \to \begin{bmatrix} R(-3)^2 \\ \oplus \\ R(-4) \end{bmatrix} \to J_1 \to 0.$$

Let $\mathbb{Z}_2$ be the residual to $\mathbb{Z}_1$ in the complete intersection $\mathbb{X}_1$ defined by J_1. What can we say about $\mathbb{Z}_2$?

1) By the Cayley-Bacharach Theorem (see [**15**]) the h-vector of $\mathbb{Z}_2$ is $(1, 3, 6, 8, 6)$.

2) There is a short exact sequence

(6.3) $$0 \to J_1 \xrightarrow{\alpha} I_{\mathbb{Z}_1} \to K_{\mathbb{Z}_2}(-6) \to 0$$

where $K_{\mathbb{Z}_2}$ is the canonical module of $\mathbb{Z}_2$ (cf. [[**56**], Lemma 5.10 and Example 5.7,(iii)]).

3) Using the short exact sequence (6.3) and the resolutions (6.1) and (6.2), we can compute the minimal free resolution of $I_{\mathbb{Z}_2}$ as follows. Consider the commutative diagram

$$\begin{array}{ccccccccc}
 & & & & & & & & 0 \\
 & & & & & & & & \uparrow \\
 & & & & & & & & K_{\mathbb{Z}_2}(-6) \\
 & & & & & & & & \uparrow \\
0 & \to & R(-6)^2 & \to & R(-4)^9 & \to & R(-3)^8 & \to & I_{\mathbb{Z}_1} & \to & 0 \\
 & & \uparrow \alpha_3 & & \uparrow \alpha_2 & & \uparrow \alpha_1 & & \uparrow \alpha \\
0 & \to & R(-10) & \to & \begin{bmatrix} R(-6) \\ \oplus \\ R(-7)^2 \end{bmatrix} & \to & \begin{bmatrix} R(-3)^2 \\ \oplus \\ R(-4) \end{bmatrix} & \to & J_1 & \to & 0 \\
 & & & & & & & & \uparrow \\
 & & & & & & & & 0
\end{array}$$

where the maps α_i are lifted from α.

Note that the two cubic generators of J_1 are also minimal generators of $I_{\mathbb{Z}_1}$. This means that the restriction of the map α_1 to two copies of $R(-3)$ is an isomorphism; we will use this fact shortly to split off some terms.

The mapping cone of the above diagram gives a free resolution (not minimal) for $K_{\mathbb{Z}_2}(-6)$:

$$0 \to R(-10) \to \begin{bmatrix} R(-6)^3 \\ \oplus \\ R(-7)^2 \end{bmatrix} \to \begin{bmatrix} R(-3)^2 \\ \oplus \\ R(-4)^{10} \end{bmatrix} \to R(-3)^8 \to K_{\mathbb{Z}_2}(-6) \to 0.$$

The observation above about α_1 means that two copies of $R(-3)$ split off, giving a (what has to be) *minimal* free resolution

$$0 \to R(-10) \to \begin{bmatrix} R(-6)^3 \\ \oplus \\ R(-7)^2 \end{bmatrix} \to R(-4)^{10} \to R(-3)^6 \to K_{\mathbb{Z}_2}(-6) \to 0.$$

Dualizing this resolution and twisting by -10 gives a minimal free resolution for $R/I_{\mathbb{Z}_2}$, so we get the minimal free resolution

$$0 \to R(-7)^6 \to R(-6)^{10} \to \begin{bmatrix} R(-3)^2 \\ \oplus \\ R(-4)^3 \end{bmatrix} \to I_{\mathbb{Z}_2} \to 0.$$

Now we choose a general regular sequence J_2 of type $(4, 4, 4)$ in $I_{\mathbb{Z}_2}$. In the same way as above, we obtain a residual set of points, $\mathbb{Z}$, with the desired h-vector, $(1, 3, 6, 10, 12, 6, 2)$, and free resolution given by the mapping cone obtained from the diagram

$$\begin{array}{ccccccccc}
0 & \to & R(-7)^6 & \to & R(-6)^{10} & \to & \begin{bmatrix} R(-3)^2 \\ \oplus \\ R(-4)^3 \end{bmatrix} & \to & I_{\mathbb{Z}_2} & \to & 0 \\
& & \uparrow & & \uparrow & & \uparrow & & \uparrow & & \\
0 & \to & R(-12) & \to & R(-8)^3 & \to & R(-4)^3 & \to & J_2 & \to & 0
\end{array}$$

and after splitting the three generators of degree 4, the mapping cone gives that the residual, $\mathbb{Z}$, has minimal free resolution

$$0 \to R(-9)^2 \to R(-6)^{10} \to \begin{bmatrix} R(-4)^3 \\ \oplus \\ R(-5)^6 \end{bmatrix} \to I_{\mathbb{Z}} \to 0$$

as desired.

The difficult part of using liaison as a method of construction is to determine in advance what object to start with (above we used 12 general points in $\mathbb{P}^3$) and what sequence of links to use in order to obtain the desired result. Usually one proceeds in reverse! One starts with the (hypothetical) desired object and tries to find a sequence of links that, if it really exists, will result in a known object (above it is a set of 12 points). Then one makes the actual computation, starting with the known object and working backwards as above, and one verifies that at the end one obtains the desired level algebra.

As described above, this is a somewhat "educated hit and miss" procedure. In our case of socle degree 6, though, we have an incredibly powerful tool at our disposal. Using the methods described earlier in this monograph, we had constructed all possible level Artinian Hilbert functions of socle degree 6 explicitly as sums of ideals of points in $\mathbb{P}^2$. Having these explicit constructions already, it is a very simple matter to study them on the computer and look for "small" links. Applying this idea enabled us to find examples of Hilbert functions of level point sets in $\mathbb{P}^3$ that we were unable to find with any other method. It also resulted in some interesting observations, noted below. First, though, we illustrate the idea.

EXAMPLE 6.11. The compressed Hilbert function obtained in Example 6.10 is obtained using the linked-sum method with points in $\mathbb{P}^2$, in the following way (in this diagram, $\mathbb{X}$ corresponds to the points $\bullet$ and $\mathbb{Y}$ corresponds to the points $*$).

$$\mathbb{Z} = \left\{ \begin{array}{cccccc}
* & \bullet & * & & & \\
* & * & * & & & \\
* & \bullet & \bullet & & & \\
\bullet & * & * & \bullet & * & * \\
\bullet & \bullet & * & \bullet & \bullet & \bullet \\
\bullet & \bullet & \bullet & * & \bullet & *
\end{array} \right\}$$

Notice that this configuration does not lend itself to the use of Method 1 since the two subsets $\mathbb{X}$ and $\mathbb{Y}$ do not satisfy the "nested" property.

The Hilbert functions of the various sets of points in the diagram are given by

$$\begin{array}{rcccccccccc}
\Delta \mathbf{H}(R/I_\mathbb{Z}, -) & : & 1 & 3 & 6 & 10 & 15 & 21 & 25 & 27 & \to \\
\Delta \mathbf{H}(R/I_\mathbb{X}, -) & : & 1 & 3 & 6 & 10 & 13 & 13 & 13 & 13 & \to \\
\Delta \mathbf{H}(R/I_\mathbb{Y}, -) & : & 1 & 3 & 6 & 10 & 14 & 14 & 14 & 14 & \to \\
\Delta \mathbf{H}(R/(I_\mathbb{X}+I_\mathbb{Y}), -) & : & 1 & 3 & 6 & 10 & 12 & 6 & 2 & 0 & \to.
\end{array}$$

The Artinian ideal $I_\mathbb{X} + I_\mathbb{Y}$ can be studied on the computer (we did it with Macaulay [**2**]), and if we choose at each step the smallest complete intersection possible, we obtain that the following sequence of links takes us from this algebra to one with h-vector $(1,3,6,2)$:

$$(4,4,5), \quad (4,4,4), \quad (3,4,4), \quad (3,3,4).$$

Now, noting that this last ideal has the Hilbert function and graded Betti numbers of the Artinian reduction of a set of 12 general points, and noting that complete intersection links of a hyperplane section (or Artinian reduction) lift (the proof of [[**53**], Proposition 5.2.25] works also in the Artinian case), we get a level set of points in $\mathbb{P}^3$ by starting with 12 general points and applying the above links in reverse order.

Note that in Example 6.10 we were able to achieve our result with two links rather than four.

REMARK 6.12. 1) Watanabe [**70**] has shown that every height three Gorenstein ideal is licci. Since Gorenstein ideals are level of type 1, it is natural to wonder if all level algebras might be licci. Example 6.10 shows that this is not the case, even for type 2. Indeed, the level algebra that we constructed is linked to a set of 12 points in $\mathbb{P}^3$ with generic Hilbert function and minimal free resolution. The defining ideal of these points has 8 minimal generators, all of degree 3. But Huneke [**42**] has shown that if an ideal I, with height g and $\nu(I)$ generators all of degree d, is licci then

$$\binom{\nu(I)+1}{2} \leq \binom{2d+g-1}{g-1}.$$

Since this inequality is violated with this example, we get that our set of points with compressed Hilbert function is not licci.

On the other hand, Hartshorne [**36**] has shown that for any n, a general set of n points on a smooth cubic surface in $\mathbb{P}^3$ is glicci. Thus, the points in Example 6.10 are at least glicci.

2) Let $\mathbb{V}$ be a subscheme of $\mathbb{P}^n$. In order to find the "smallest" subscheme in the liaison class of $\mathbb{V}$ (or to a complete intersection in case $\mathbb{V}$ is licci), a very natural procedure to try is the one given above, namely at each step choose the *smallest* complete intersection containing the subscheme. It is known in codimension two that this works, and is the most efficient method, for arithmetically Cohen-Macaulay subschemes.

We were quite surprised to see, as a result of our calculations, that this method is sometimes not the best, and indeed sometimes does not work, in higher codimension.

The first illustration is given above, where the set of points with compressed Hilbert function can be linked to 12 general points in two steps, but the "minimal link" procedure requires four steps.

But a more interesting example is the following. The h-vector $(1, 3, 6, 8, 7, 6, 2)$ can be constructed using the linked sum method for points in $\mathbb{P}^2$ as follows: in the diagram below take $\mathbb{X}$ as the points represented by •'s and $\mathbb{Y}$ as the points represented by $*$'s and let $J = I_\mathbb{X} + I_\mathbb{Y}$. From (5.16) above, the set $\mathbb{Z}$ is a level set of points in $\mathbb{P}^2$ with socle degree 7 and type 2.

$$\mathbb{Z} = \left\{ \begin{array}{cccccc} \bullet & * & * & * & & \\ \bullet & \bullet & \bullet & \bullet & & \\ * & * & * & \bullet & * & * \\ * & * & * & * & * & \bullet \\ * & * & * & \bullet & * & * \end{array} \right\}.$$

The Hilbert functions of the various sets of points involved are tabulated below:

$$\begin{array}{llcccccccc}
\Delta\mathbf{H}(R/I_\mathbb{Z}, -) & : & 1 & 3 & 6 & 10 & 15 & 20 & 24 & 26 & \to \\
\Delta\mathbf{H}(R/I_\mathbb{X}, -) & : & 1 & 3 & 6 & 8 & 8 & 8 & 8 & 8 & \to \\
\Delta\mathbf{H}(R/I_\mathbb{Y}, -) & : & 1 & 3 & 6 & 10 & 14 & 18 & 18 & 18 & \to \\
\Delta\mathbf{H}(R/I_\mathbb{X} + I_\mathbb{Y}, -) & : & 1 & 3 & 6 & 8 & 7 & 6 & 2 & 0 & \to.
\end{array}$$

One can check, by explicitly forming the ideal $J = I_\mathbb{X} + I_\mathbb{Y}$, that the least degrees giving a regular sequence in it are $(3, 4, 6)$, and choosing such a complete intersection links J to some residual J'. But the smallest link for J' is again of type $(3, 4, 6)$, so we link back to an ideal with the same Hilbert function and graded Betti numbers as J. On the other hand, if one starts instead with the link $(4, 4, 6)$ (which is *not* the smallest possible), the sequence of links

$$(4, 4, 6), \ (4, 4, 5), \ (2, 2, 5), \ (1, 2, 2)$$

result in an ideal with h-vector (1). This can be lifted to a point in $\mathbb{P}^3$, and then this sequence of links can be used backwards, to produce a level set of points $\mathbb{V}$ with the desired h-vector, which is thus obviously licci but can not be linked to a complete intersection via minimal links.

3) The method above consisted of explicitly constructing an Artinian level algebra in $k[x_1, x_2, x_3]$, following the complete intersections that link it to a simple Artinian algebra, lifting this to a set of points $\mathbb{Z}_1$, and then reversing the sequence of numerical links to obtain our desired set of points $\mathbb{Z}$ with the right Hilbert function and resolution. Note that since links lift, if the "simple Artinian algebra" is actually the Artinian reduction of $\mathbb{Z}_1$ (as opposed to simply having the same numerical data), then working backwards we can construct a set of points whose Artinian reduction is actually the original algebra, rather than simply having the right Hilbert function and graded Betti numbers. This is a much stronger conclusion!

6.3. Method 3: Union of suitable sets of points

Another natural way to try to produce sets of points in $\mathbb{P}^3$ is as a union of smaller sets. This is done, for instance, to produce the pure configurations in $\mathbb{P}^2$ that are used to make sums of ideals of points and construct our height 3 level algebras. We believe that there should be many such constructions, but here we give just one (and a corollary).

LEMMA 6.13. *Let $\mathbb{Z}_1 \subset \mathbb{P}^3$ be arithmetically Gorenstein with socle degree s. Let $\mathcal{C} \subset \mathbb{P}^3$ be a complete intersection curve containing $\mathbb{Z}_1$ (as schemes), with $I_\mathcal{C} = (F_1, F_2)$ and $\deg F_1 = d$, $\deg F_2 = e$. Let F be a general polynomial of degree f. Assume that $d + e = s + 3$. Then the ideal*

$$I_\mathbb{Z} = F \cdot I_{\mathbb{Z}_1} + I_\mathcal{C}$$

is the saturated ideal of a scheme $\mathbb{Z}$ which is level of type two and has socle degree $s + f$.

PROOF. This is a basic double link. See [**27**] or [**48**] for a more general version that can be used in higher codimension. This gives that the ideal is saturated. We have, from [[**48**], Lemma 4.8 (in the proof)], that there is an exact sequence

$$0 \to I_\mathcal{C}(-f) \to I_{\mathbb{Z}_1}(-f) \oplus I_\mathcal{C} \to I_\mathbb{Z} \to 0.$$

Suppose that we have a minimal free resolution

$$0 \to R(-s-3) \to \mathbb{F}_2 \to \mathbb{F}_1 \to I_{\mathbb{Z}_1} \to 0$$

for $I_{\mathbb{Z}_1}$. Then we have

$$\begin{array}{ccccccccc}
 & & & & & & 0 & & \\
 & & & & & & \downarrow & & \\
 & & 0 & & & & R(-s-3-f) \oplus 0 & & \\
 & & \downarrow & & & & \downarrow & & \\
 & & R(-d-e-f) & & & & \mathbb{F}_2(-f) \oplus R(-d-e) & & \\
 & & \downarrow & & & & \downarrow & & \\
 & & R(-d-f) \oplus R(-e-f) & & & & \mathbb{F}_1(-f) \oplus R(-d) \oplus R(-e) & & \\
 & & \downarrow & & & & \downarrow & & \\
0 & \to & I_\mathcal{C}(-f) & \to & & & I_\mathbb{Z}(-f) \oplus I_\mathcal{C} & \to I_\mathbb{Z} \to 0 \\
 & & \downarrow & & & & \downarrow & & \\
 & & 0 & & & & 0 & &
\end{array}$$

A mapping cone gives a free resolution for $I_\mathbb{Z}$, and the assumption that $d+e = s+3$ gives the desired end of the resolution and in particular that $\mathbb{Z}$ is level. Some terms may split off (for instance if the generators of $I_\mathcal{C}$ are minimal generators of $I_{\mathbb{Z}_1}$), but neither summand from the end of the resolution can split off, for numerical reasons (using the fact that $I_{\mathbb{Z}_1}$ is Gorenstein). Hence we have type two. □

We can get a union of two complete intersections to be level of type two as follows.

COROLLARY 6.14. *Let $\mathbb{Z}_1 \subset \mathbb{P}^3$ be a complete intersection of type (a, b, c) and let $\mathcal{C} \subset \mathbb{P}^3$ be a complete intersection curve of type (d, e) containing $\mathbb{Z}_1$. (Note that $\mathbb{Z}_1$ need not be a hypersurface section of $\mathcal{C}$.) Let $\mathbb{Z}_2$ be a degree f hypersurface section of $\mathcal{C}$. Assume that $\mathbb{Z}_1$ and $\mathbb{Z}_2$ have no common component.*

If $a + b + c = d + e$ then $\mathbb{Z} := \mathbb{Z}_1 \bigcup \mathbb{Z}_2$ is level of type two and socle degree $a + b + c + f - 3 = d + e + f - 3$.

REMARK 6.15. Corollary 6.14 can easily be extended to higher socle type. Suppose that $\mathbb{Z}_1$ is level of arbitrary type (rather than being of type 1 as in the Corollary). Then the mapping cone in the proof of Lemma 6.13 adds one free summand to the last free module of the minimal free resolution of $\mathbb{Z}_1$, and by suitable choice of $\mathcal{C}$ we can guarantee that the resulting set $\mathbb{Z}$ is again level.

Notice that in order to repeat the procedure indefinitely, the curve $\mathcal{C}$ has to get progressively bigger. However, if we fix the desired socle type, we can also view the procedure in reverse. From this point of view, we can give a codimension three version of the "pure configuration" construction that is at the heart of the procedure that gives most of the Artinian level algebras constructed in this monograph. Furthermore, this approach gives an alternative way to view (and prove) the results even in the codimension two case. We first consider the latter case from this point of view. Note that in order to be consistent with the notation of the earlier parts of this monograph, we must also reverse the roles of $\mathbb{Z}_1$ and $\mathbb{Z}_2$ in Corollary 6.14.

The pure configuration begins with a complete intersection $\mathbb{Z}_1$ in $\mathbb{P}^2$, namely a box of size $a \times b$ (let us think of a as giving the number of columns, i.e. the width of the box, and b as giving the number of rows, i.e. the height of the box). Then we add a second complete intersection $\mathbb{Z}_2$ to it, namely a box of size $c \times d$ subject to the conditions that $b > d$ (i.e. the new box is "shorter" than the first one) and $c + d = b$. Note that the value of a is not part of the hypotheses. The union of these two boxes is level of type 2 and socle degree $a + b - 2$. (Note that $a + b$ is the sum of the dimensions of the first box.) This procedure can be repeated to produce level algebras of higher socle type, which is the number of boxes.

In fact, $\mathbb{Z}_1$ is obtained as a hypersurface section of a curve containing $\mathbb{Z}_2$: one can think of the b horizontal lines on which $\mathbb{Z}_1$ lies as being the curve and the a vertical lines as giving the hypersurface section. The condition that $b > d$ is just so that the curve of horizontal lines contains $\mathbb{Z}_2$.

Corollary 6.14 gives a codimension three version of this construction. We start with a box $\mathbb{Z}_1$, (now three dimensional) of size $a \times b \times c$. We then add a box of size $d \times e \times f$, $\mathbb{Z}_2$. The constraints are that the complete intersection curve of type (a,b) should contain $\mathbb{Z}_2$, which can be translated to a set of inequalities, and that $d + e + f = a + b$. Note that the value of c is not part of the hypotheses.

Again, this can be extended to higher socle type.

6.4. Method 4: General Sets of Points on Suitable curves

Sometimes we can find level sets of points by choosing appropriate (sometimes general) points on a suitable curve, or simply with an appropriate configuration (e.g. chosen generically in $\mathbb{P}^3$). To prove that this set of points has the desired property, it can be useful to apply some of the above techniques. We illustrate with an example, but see also Chapter 7.

EXAMPLE 6.16. Consider the h-vector $(1, 3, 5, 5, 5, 5, 2)$. Through degree 5 this agrees with the first difference of the Hilbert function of an arithmetically Cohen-Macaulay curve $\mathcal{C}$ of degree 5 and genus 2 in $\mathbb{P}^3$. This strongly suggests that a general choice, $\mathbb{Z}$, of 26 points on $\mathcal{C}$ will have the desired properties. This is not hard to verify on the computer, but it can be checked directly.

$\mathbb{Z}$ can clearly be linked to something using a complete intersection of type $(2, 3, 6)$, but to what? The first two generators link $\mathcal{C}$ to a line, and the whole complete intersection then links $\mathbb{Z}$ to a set $\mathbb{Z}_1$ with h-vector $(1, 3, 3, 1, 1, 1)$ consisting of 6 points on that line plus four other points. This set can then clearly be linked in a complete intersection of type $(2, 2, 6)$ to get a residual with h-vector $(1, 3, 3, 3, 3, 1)$. This is the h-vector of a set of 14 points on a twisted cubic curve, and such a set of points is arithmetically Gorenstein.

So we are led to the following. Let $\mathcal{C}'$ be a twisted cubic curve in $\mathbb{P}^3$ containing a point P. Let $\mathbb{Z}_1$ be an element of the linear system $5H - P$ on $\mathcal{C}'$. It is not hard to check that $\mathbb{Z}_1$ is arithmetically Gorenstein with h-vector $(1,3,3,3,3,1)$ and minimal free resolution

$$0 \to R(-8) \to \begin{array}{c} R(-3)^2 \\ \oplus \\ R(-6)^3 \end{array} \to \begin{array}{c} R(-2)^3 \\ \oplus \\ R(-5)^2 \end{array} \to I_{\mathbb{Z}_1} \to 0.$$

Then applying, successively, links of type $(2,2,6)$ and $(2,3,6)$ and using the mapping cone construction (with the corresponding splitting off of generators), we verify the desired Hilbert function, and minimal free resolution is

$$0 \to R(-9)^2 \to \begin{array}{c} R(-4)^2 \\ \oplus \\ R(-7)^4 \\ \oplus \\ R(-8) \end{array} \to \begin{array}{c} R(-2) \\ \oplus \\ R(-3)^2 \\ \oplus \\ R(-6)^3 \end{array} \to I_\mathbb{Z} \to 0.$$

In cases where we have level point sets on curves it is sometimes possible to verify the WLP for an Artinian reduction of their homogeneous coordinate ring. For example, we have the following proposition.

PROPOSITION 6.17. *Let $\mathcal{C} \subset \mathbb{P}^n$ be a reduced arithmetically Cohen Macaulay curve. Let $\mathbb{X} \subset \mathcal{C}$ be a set of points. Let $R = k[x_0, \ldots, x_n]$, $T = k[x_1, \ldots, x_n]$ (where we can assume, with no loss of generality, that no points of $\mathbb{X}$ nor any component of $\mathcal{C}$ lies in the hyperplane defined by x_0 - so x_0 is not a zero-divisor on either $R/I_\mathbb{X}$ or $R/I_\mathcal{C}$). Let $J = (I_\mathbb{X} + (x_0))/(x_0) \subset T$ be the ideal defining the Artinian reduction of $R/I_\mathbb{X}$. Let d be the last degree for which $(I_\mathbb{X})_d = (I_\mathcal{C})_d$ and assume that $\deg \mathcal{C} = \Delta \mathbf{H}(\mathcal{C}, d)$.*

Then T/J has the WLP.

PROOF. Since $\mathbb{X} \subset \mathcal{C}$ we have a surjection $R/I_\mathcal{C} \to R/I_\mathbb{X}$. Furthermore, if we let $I = (I_\mathcal{C} + (x_0))/(x_0)$, we also have a surjection

$$\eta : T/I \longrightarrow T/J.$$

Note that T/I is a Cohen-Macaulay ring which is the homogeneous coordinate ring of the points of the hyperplane section of $\mathcal{C}$ given by x_0. Since T/J agrees with T/I in all degrees $\leq d$, it follows that for a general linear form L, the induced map $\times L : (T/J)_i \to (T/J)_{i+1}$ is an injection, for all $i < d$. We also know, from the fact that $\Delta \mathbf{H}(\mathcal{C}, d) = \deg \mathcal{C}$, that $(T/I)_i \xrightarrow{\times L} (T/I)_{i+1}$ is an isomorphism for all $i \geq d$.

We have to show that for a general linear form in T the multiplication map,

$$(T/J)_i \to (T/J)_{i+1}$$

is a surjection, for all $i \geq d$. For any linear form, L, in T_1 we have a commutative diagram:

$$\begin{array}{ccc} (T/I)_i & \xrightarrow{\times L} & (T/I)_{i+1} \\ \eta_i \downarrow & & \downarrow \eta_{i+1} \\ (T/J)_i & \xrightarrow{\times L} & (T/J)_{i+1}. \end{array}$$

Assume that L is sufficiently general. As we noted above, the top row is an isomorphism, and the vertical maps are surjective. It follows that the bottom row is a surjection, as claimed. □

EXAMPLE 6.18. Let $h = (1, 3, 6, 9, 11, 4)$. It is possible to show (we did this on a computer) that this is the h-vector of a level set, $\mathbb{X}$, of 34 points on a smooth arithmetically Cohen-Macaulay curve, $\mathcal{C}$, of degree 11 and genus 15 that is linked to a line.

The curve $\mathcal{C}$ has Hilbert function

$$\mathbf{H}_\mathcal{C} \; : \; 1 \quad 4 \quad 10 \quad 19 \quad 30 \quad 41 \; \overset{+11}{\longrightarrow}$$

and the set of points $\mathbb{X}$ has Hilbert function

$$\mathbf{H}_\mathbb{X} \; : \; 1 \quad 4 \quad 10 \quad 19 \quad 30 \quad 34 \; \longrightarrow .$$

Now, since $I_\mathbb{X} = I_\mathcal{C}$ for degrees ≤ 4 and $\mathcal{C}$ is ACM, multiplication by a general linear form on the Artinian reduction of the homogeneous coordinate ring of $\mathbb{X}$ (call it B) is an injection up to degree 4. By Proposition 6.17 we have that the multiplication map of a general linear form, from B_4 to B_5 is surjective. Thus, B is a level Artinian algebra with the WLP.

CHAPTER 7

Expected Behavior

In this chapter we discuss two situations of "expected" behavior of an ideal, and the connections to level algebras.

First recall that Hochster and Laksov [40] proved that in a polynomial ring $R = k[x_1, \ldots, x_n]$, a general set of polynomials $F_1, \ldots, F_r$ of the same degree, d, will span the maximum dimension possible in degree $d+1$. That is, the dimension of the image of the multiplication map

$$\langle F_1, \ldots, F_r \rangle \otimes R_1 \to R_{d+1}$$

is $\min\{n \cdot r, \dim R_{d+1}\}$ (the map is either injective or surjective).

Now let $\mathbb{Z}$ be a non-degenerate set of points in $\mathbb{P}^{n-1}$ and let d be the degree in which the Hilbert function of $\mathbb{Z}$ achieves the multiplicity of $\mathbb{Z}$ (or any degree past this point), so $\mathbf{H}(\mathbb{Z}, d) = \mathbf{H}(\mathbb{Z}, d+1) = |\mathbb{Z}|$. Let r be an integer such that $r < |\mathbb{Z}|$. A general set of r polynomials $F_1, \ldots, F_r$ of degree d will all be non-zero, and in fact will be linearly independent, in $(R/I_\mathbb{Z})_d$. It is natural to ask under what circumstances the type of behavior described by Hochster and Laksov continues to hold in $R/I_\mathbb{Z}$. That is, when is it true that the image of the map

$$\langle F_1, \ldots, F_r \rangle \otimes (R/I_\mathbb{Z})_1 \to (R/I_\mathbb{Z})_{d+1}$$

has the "expected" dimension $\min\{r \cdot n, \dim(R/I_\mathbb{Z})_{d+1}\}$. Notice that we are forming the Artinian algebra $R/(I_\mathbb{Z} + (F_1, \ldots, F_r))$ and asking for its Hilbert function.

Liaison provides an interesting connection between this question and the question of the existence of level algebras with certain Hilbert functions. Indeed, it often turns out that "expected" ideal generation behavior is **linked** (in both senses of the word!!) to "expected" canonical module generation, which in turn often means level algebras.

The connection between these questions is reflected in the proof of the following result. The statement focuses on the existence of level algebras, and gives a surprising geometric condition for this existence. For convenience we assume that we have only three variables.

PROPOSITION 7.1. *Consider the sequence* $\underline{h} = (1, 3, \ldots, h_{s-1}, h_s)$, *with* $h_s \leq h_{s-1}$, *and* $3 \cdot h_s \geq h_{s-1}$. *Assume that*

$$(1, 3, \ldots, h_{s-1}, h_{s-1}, \ldots)$$

is the Hilbert function of a non-degenerate set of h_{s-1} *points,* $\mathbb{Z}$, *in* $\mathbb{P}^2$.

Let $V = \langle F_1, \ldots, F_{h_s} \rangle$ *be the vector space spanned by a general set of* h_s *forms of degree* s. *Assume that* $\mathbb{Z}$ *imposes independent conditions on the linear system* $|R_1 \cdot V|$ *of hypersurfaces in projective space. Then*

(a) $R/(I_\mathbb{Z} + (F_1, \ldots, F_{h_s}))$ *has the expected Hilbert function*

$$(1, 3, \ldots, h_{s-1}, (h_{s-1} - h_s), 0).$$

67

(b) $\underline{h}$ is a level sequence.

(c) There exists an example of an Artinian level algebra with Hilbert function $\underline{h}$, with WLP, that is a quotient of $R/I_\mathbb{Z}$.

Conversely, if $\mathbb{Z}$ does not impose independent conditions on $|R_1 \cdot V|$ then $R/(I_\mathbb{Z} + (F_1, \ldots, F_{h_s}))$ does not have the expected Hilbert function. Furthermore, no Artinian quotient of $R/I_\mathbb{Z}$ is level with Hilbert function $\underline{h}$.

PROOF. We will consider ideals of the form $J = I_\mathbb{Z} + (F_1, \ldots, F_k)$ for some suitable polynomials $F_1, \ldots, F_k$ of degree s. (We are not claiming that all level algebras with Hilbert function $\underline{h}$ come from ideals of this form.) We first check that if J is of this form then R/J has the WLP (whether or not it is level or has Hilbert function $\underline{h}$). The argument is almost identical to that in Proposition 6.17. Indeed, let L be a general linear form. R/J agrees with $R/I_\mathbb{Z}$ in degrees $\leq s - 1$, so $\times L : (R/J)_i \to (R/J)_{i+1}$ is clearly injective for $i \leq s - 2$. When $i \geq s - 1$ we have a commutative diagram

$$\begin{array}{ccc} (R/I_\mathbb{Z})_i & \xrightarrow{\times L} & (R/I_\mathbb{Z})_{i+1} \\ \downarrow & & \downarrow \\ (R/J)_i & \xrightarrow{\times L} & (R/J)_{i+1} \\ \downarrow & & \downarrow \\ 0 & & 0 \end{array}$$

By our hypotheses, the top horizontal map is an isomorphism, so $\times L : (R/J)_i \to (R/J)_{i+1}$ is surjective, as desired.

Suppose now that $\mathbb{Z}$ imposes independent conditions on $|R_1 \cdot V|$, where $V = \langle F_1, \ldots, F_{h_s} \rangle$ is the vector space spanned by a sufficiently general set of h_s polynomials of degree s. Thanks to the assumption that $3 \cdot h_s \geq h_{s-1}$, and the theorem of Hochster and Laksov [**40**], we have that the component R_{s+1} in degree $s+1$ of the ideal $(F_1, \ldots, F_{h_s})$ has dimension $\geq h_{s-1}$. That is, $\dim_k(R_1 \cdot V) \geq h_{s-1} = |\mathbb{Z}| = \operatorname{codim}_{R_{s+1}}((I_\mathbb{Z})_{s+1})$ (where $|\mathbb{Z}|$ is the number of points of $\mathbb{Z}$). So it is numerically possible for the points of $\mathbb{Z}$ to impose independent conditions on $|R_1 \cdot V|$.

We now claim that our assumption that $\mathbb{Z}$ does impose independent conditions implies that the component of $R/J = R/[I_\mathbb{Z} + (F_1, \ldots, F_{h_s})] = R/[I_\mathbb{Z} + V]$ in degree $s+1$ has dimension 0. (Note that $(I_\mathbb{Z} + V)_{s+1} = (R_1 \cdot V) + (I_\mathbb{Z})_{s+1}$.) This is because independent conditions means that $\dim[(R_1 \cdot V) \cap (I_\mathbb{Z})_{s+1}] = \dim(R_1 \cdot V)_{s+1} - |\mathbb{Z}|$, so

$$\begin{aligned} \dim(I_\mathbb{Z} + V)_{s+1} &= \dim(R_1 \cdot V) + \dim(I_\mathbb{Z})_{s+1} - \dim((R_1 \cdot V) \cap I_\mathbb{Z})_{s+1} \\ &= \dim(R_1 \cdot V) + (\dim R_{s+1} - |\mathbb{Z}|) - (\dim(R_1 \cdot V) - |\mathbb{Z}|) \\ &= \dim R_{s+1} \end{aligned}$$

So under our assumptions, we have that the Hilbert function of R/J is

(7.1) $\qquad\qquad (1, 3, h_2, \ldots, h_{s-1}, (h_{s-1} - h_s), 0)$

proving (a). (What is at issue is the 0 in degree $s + 1$.) Note that without the assumption on $\mathbb{Z}$ this is not true!! This proves the first part of the converse. See Example 7.3.

Let X be a sufficiently general complete intersection containing $\mathbb{Z}$, generated by forms of degree $s + 1$ and $s + 2$ (remember that $\mathbb{Z}$ has codimension two). Let $\mathbb{Z}'$ be the residual to $\mathbb{Z}$ in X. Since $\mathbb{Z}$ is reduced and has regularity s, $\mathbb{Z}'$ has no component in common with $\mathbb{Z}$. One can check with standard liaison methods that the first generator of $I_{Z'}$ comes in degree $s+1$, and this generator comes from X. The first additional generator comes in degree $s+2$.

Let $I' = I_Z + I_{Z'}$. Note that R/I' is Gorenstein with Hilbert function

$$(1, 3, h_2, \ldots, h_{s-2}, h_{s-1}, h_{s-1}, h_{s-1}, h_{s-2}, \ldots, h_2, 3, 1).$$

Note also that in degrees $\leq s+1$, I' agrees with I_Z. Let $J = I' + (F_1, \ldots, F_{h_s})$. Then we have

$$I' \subset J := I' + (F_1, \ldots, F_{h_s}) = I_Z + (F_1, \ldots, F_{h_s})$$

and R/J has Hilbert function (7.1). Consequently, by the theorem of Davis, Geramita and Orecchia [**15**], the residual to J in I' has Hilbert function $\underline{h}$. Let us denote by J' this ideal. We claim that R/J' is level, and since it has the desired Hilbert function, we will be done with (b).

To see this, we again turn to the mapping cone. We want to show that the end of the minimal free resolution of R/J' has only one twist. Suppose that I_Z has minimal free resolution

$$0 \to \mathbb{G}_2 \to \mathbb{G}_1 \to I_Z \to 0$$

where all the summands in $\mathbb{G}_1$ are of the form $R(-t)$ for $t \leq s$. Then J has minimal free resolution of the form

$$\cdots \to \mathbb{G}_1 \oplus R(-s)^{h_s} \to J \to 0$$

and I' has minimal free resolution

$$\cdots \to \mathbb{G}_1 \oplus \mathbb{F}_1 \to I' \to 0$$

where every summand of $\mathbb{F}_1$ is of the form $R(-t)$ for some $t \geq s+2$ (but it is not necessarily the case that all summands have the same value of t).

We then have a diagram

$$\begin{array}{ccccc} \cdots & \to & \mathbb{G}_1 \oplus R(-s)^{h_s} & \to & J \to 0 \\ & & \uparrow & & \uparrow \\ \cdots & \to & \mathbb{G}_1 \oplus \mathbb{F}_1 & \to & I' \to 0 \end{array}$$

It is clear that the summands of $\mathbb{G}_1$ split after taking the mapping cone, and no summand of $\mathbb{F}_1$ is of the form $R(-s)$, so the mapping cone gives the residual J' with a minimal free resolution that ends with a twist of R^{h_s}, i.e. R/J' is level.

The proof that R/J' has WLP is almost identical to the argument at the beginning of this proof (but replacing R/I_Z with R/I') and is omitted.

Finally, we have to prove that if J is an Artinian ideal containing I_Z, with Hilbert function $\underline{h}$, then R/J is not level. Suppose otherwise. We know that $J = I_Z + \langle G_1, \ldots, G_k \rangle$, where the G_i include at least $h_{s-1} - h_s$ generators of degree s, and possibly some generators of degree $s+1$. But in a completely analogous way to the construction of I' above, we can use liaison to find a Gorenstein ideal I' such that $J = I' + \langle G_1, \ldots, G_k \rangle$, and so link J using I' to a residual J'. The fact that R/J is level with Hilbert function $\underline{h}$ implies that J' is of the form $I' + \langle F_1, \ldots, F_{h_s} \rangle$ (from the mapping cone) with Hilbert function $(1, 3, h_2, \ldots, h_{s-1}, (h_{s-1} - h_s), 0)$. But we have just seen that R/J' can *not* have this expected Hilbert function. Contradiction. □

REMARK 7.2. The above argument shows that the residual to the ideal $I_Z + (F_1, \ldots, F_{h_s})$ is level in any case, as long as it is Artinian, and it is not hard to see that this will always be true under our hypotheses. However, it might not have the right Hilbert function.

EXAMPLE 7.3. As w will see in the appendices, one of the few Hilbert functions of low socle degree that is not immediately covered by the methods and results of the previous chapters is the sequence

$$1\ 3\ 4\ 5\ 6\ 2.$$

It was shown by Cho and Iarrobino [**14**] that this is not a level sequence. In fact, they showed that any sequence $(1, 3, 4, 5, 6, \ldots, 2)$, where the missing entries are all 6's, is not a level sequence. In Example 5.9 we ruled out a similar sequence using inverse systems, and on page 80 we will use inverse systems to show that $(1, 3, 4, 5, 6, 6, 2)$ is also not a level sequence. The purpose of this example is to complement the result of Cho and Iarrobino with a geometric interpretation, not to re-prove their result. In fact, this example will also show why we can't hope for a result like the conclusion of Proposition 7.1 without the assumption on imposing independent conditions.

Consider the sequence $\underline{h} = (1, 3, 4, 5, 6, \ldots, 2)$ as above, and say that the socle degree is $s \geq 5$. The sequence $(1, 3, 4, 5, 6, 6, \ldots)$ is a differentiable O-sequence corresponding to a zeroscheme $\mathbb{Z}$ of degree 6. Such a zeroscheme has saturated ideal (LL_1, LL_2, F) for some F of degree 5, where clearly L_2 is not a scalar multiple of L_1. We have the following properties for this ideal:

a. We can assume without loss of generality that L_2 is not a scalar multiple of L. If L_1 is a scalar multiple of L then $\mathbb{Z}$ is not reduced.
b. Let P be the point defined by $\langle L_1, L_2 \rangle$. Then F vanishes at P.
c. $\mathbb{Z}$ has a subscheme of degree 5 that lies on the line L. This subscheme is the complete intersection of F and L.

We want to explore the question of whether a quotient of $R/I_\mathbb{Z}$ with Hilbert function $\underline{h}$ can be level. Then we will explore the question of whether we can always reduce to this situation.

We have $s \geq 5$, $h_s = 2$. Let $V = (F_1, F_2)$ be the vector space spanned by two forms of degree s that are not zero in $R/I_\mathbb{Z}$. We saw in Proposition 7.1 that if $\mathbb{Z}$ does not impose independent conditions on $|R_1 \cdot V|$ then no quotient of $R/I_\mathbb{Z}$ is level with Hilbert function $\underline{h}$. But $\mathbb{Z}$ imposes independent conditions if and only if $\dim(R_1 \cdot V) \cap (I_\mathbb{Z})_6 = 0$. This, in turn, is true if and only if there is no non-trivial relation of the form

$$A_1(LL_1) + A_2(LL_2) + A_3 F + A_4 F_1 + A_5 F_2 = 0$$

with $\deg A_1 = \deg A_2 = 4$ and $\deg A_i = 1$ for $i \geq 3$. But since (F_1, F_2) is a pencil, there is some scalar linear combination $\lambda_1 F_1 + \lambda_2 F_2$ that vanishes on P. So $\lambda_1 LF_1 + \lambda_2 LF_2$ is in $I_\mathbb{Z}$, and this provides the non-trivial syzygy. So no ideal of the form $I_\mathbb{Z} + \langle F_1, F_2 \rangle$ has Hilbert function $(1, 3, 4, 5, 6, \ldots, 4, 0)$. Consequently, no quotient of $R/I_\mathbb{Z}$ with Hilbert $(1, 3, 4, 5, 6, \ldots, 2)$ is level.

Finally, we briefly investigate to what extent this analysis can be used to show that no level algebra exists with Hilbert function $\underline{h}$. Suppose that $A = R/I$ is any Artinian level algebra with Hilbert function $(1, 3, 4, 5, 6, \ldots, 2)$ (where the dots represent only 6's, possibly none).

Case 1: Suppose that the desired level sequence is $(1, 3, 4, 5, 6, 2)$ (i.e. $s = 5$). Consider the ideal $\langle I_{\leq 4} \rangle$ generated by the components of I of degrees ≤ 4. Because of the maximal growth in degrees 2, 3, and 4, $\langle I_{\leq 4} \rangle$ has a linear GCD in degrees 2, 3 and 4, and no further generator. Hence it is of the form $\langle LL_1, LL_2 \rangle$, where L_2 may be a scalar multiple of L. Then there exists a zeroscheme $\mathbb{Z}$ in $\mathbb{P}^2$ whose ideal agrees with I in degrees ≤ 4. Let P be the point defined by the ideal $\langle L_1, L_2 \rangle$.

Because of the maximal growth condition, it follows that I has five minimal generators in degree 5. The general element in I_5 does not have a component in common with either generator of I of degree 2. Now, inside the linear system $|I_5|$ choose a general element F with the condition that F vanishes at P. Then the ideal $\langle I_2, F \rangle$ is the saturated ideal of a zeroscheme of degree 6 with Hilbert function $(1, 3, 4, 5, 6, 6, \dots)$, and the analysis above holds to show that A is not level.

Case 2: Suppose that the desired level sequence is $(1, 3, 4, 5, 6, 6, 2)$ (resp. $(1, 3, 4, 5, 6, 6, 6, 2)$). Then there is one generator in degree 5 (resp. one generator in degree 5 and possibly one in degree 6), and the number of generators in degree 6 (resp. degree 7) depends on what this generator was. The analysis is tedious. We refer to page 80, and of course to [**14**], for a different argument that at least the first one is not a level sequence.

Case 3: Suppose that the desired level sequence is $(1, 3, 4, 5, 6, 6, 6, 6 \dots, 2)$ where again the dots represent only 6's (possibly none). Now maximal growth forces $\langle I_{\leq 7} \rangle$ to be the saturated ideal of a degree 6 subscheme of $\mathbb{P}^2$, and the above geometric analysis again applies.

We would also like to make a remark related to Section 6.4, which is the analog of the discussion above, but to produce level point sets instead of level Artinian algebras. Again for simplicity we consider the case of codimension three, but similar discussions could be made in any codimension.

Let $C \subset \mathbb{P}^3$ be a reduced curve and let $\mathbb{Z} \subset C$ be "sufficiently many" points on C, so that the Hilbert function of $\mathbb{Z}$ agrees with that of C in sufficiently large degree. Consider the minimal free resolution of I_C. If C is not arithmetically Cohen-Macaulay, this minimal free resolution has length three rather than two:

$$0 \to \mathbb{F}_3 \to \mathbb{F}_2 \to \mathbb{F}_1 \to I_C \to 0.$$

If $\mathbb{Z}$ consists of "enough" points, there is a non-zero summand of $\mathbb{F}_3$ that also corresponds to a second syzygy for $I_\mathbb{Z}$, since up to the appropriate degree $I_\mathbb{Z}$ coincides with I_C. This explains why, when looking for level sets of points on curves, it makes sense to focus on arithmetically Cohen-Macaulay curves!

Let $C \subset \mathbb{P}^3$ be an *integral* arithmetically Cohen-Macaulay curve with regularity of $\mathcal{I}_C = s$. The first difference of the Hilbert function of C has the form

(7.2) $\qquad (1, 3, h_2, \dots, h_{s-2}, h_{s-1}, h_{s-1}, \dots)$

where $\deg C = h_{s-1}$. (Note the similarity to the above situation.) Choose a set $\mathbb{Z}$ consisting of sufficiently many points on C, so that the first difference of the Hilbert function of $\mathbb{Z}$ (i.e. its h-vector) is

(7.3) $\qquad (1, 3, h_2, \dots, h_{s-2}, h_{s-1}, \dots, h_{s-1}, a, 0).$

The reverse of this sequence is the Hilbert function of the canonical module of the Artinian reduction of $R/I_\mathbb{Z}$. If $3 \cdot a \geq h_{s-1}$ (compare with the hypothesis of Proposition 7.1), we can numerically hope that the canonical module will be generated in the least degree, which means that $\mathbb{Z}$ is level. The type of obstruction evidenced in Proposition 7.1 should disappear thanks to the assumption that C is integral. We thus make the following conjecture:

CONJECTURE 7.4. *Let C be an integral arithmetically Cohen-Macaulay curve whose Hilbert function has first difference (7.2), and let $\mathbb{Z}$ be a general set of points on C with h-vector (7.3). Assume that $3 \cdot a \geq h_{s-1}$. Then $\mathbb{Z}$ is level.*

This is really a part of the so-called Minimal Resolution Conjecture for points on a Curve. Mustata [**59**] has studied this conjecture and proven it in certain cases, but in private conversation with the third author he notes that his results do not apply to this case. An example where this conjecture, if true, would apply is given on page 92, where in fact we only verified it on the computer.

The reader will note that some examples in the appendix come from a general set of points on a suitable (hyper)surface in $\mathbb{P}^3$ rather than an ACM curve in $\mathbb{P}^3$ (see for example 53] on page 107). We expect a very rich theory to emerge in analogy with the results related to the Minimal Resolution Conjecture of Lorenzini [**50**] that have emerged in recent years. Above we have focused on curves since the best work to date is in that context (as mentioned). However, we make the following observations and questions.

The Minimal Resolution Conjecture is known to hold in small projective spaces ([**22**], [**1**]), and for "many" general points in any projective space ([**69**], [**39**]). It is also known that the end of the resolution is the correct one for any number of general points in any projective space ([**49**]), and it is still open whether the same holds for the beginning of the resolution (the so-called "Ideal Generation Conjecture"). But it is known that in any projective spaces of dimension ≥ 6 (except possibly $\mathbb{P}^9$), the Minimal Resolution Conjecture fails "in the middle" of the resolution ([**18**]).

With this in mind, it is perhaps not likely that the Minimal Resolution Conjecture would hold for points on an arbitrary (or even general) ACM variety of arbitrary dimension. But is there a bound, d, such that for a "general" ACM variety, $\mathbb{X}$, of fixed Hilbert polynomial, of dimension $\leq d$, a general set of points on $\mathbb{X}$ has the expected resolution? We note that by "expected resolution" we mean the sort described by Mustata, where the "new" syzygies come only in the bottom two rows of the Betti diagram, and there are no "ghost" (i.e. redundant) terms in the resolution.

Since we know the situation for projective space, we ask the following specific question:

QUESTION 7.5. *If $\mathbb{X}$ is an integral ACM variety of dimension ≤ 5, and if $\mathbb{Z}$ is a general set of points on $\mathbb{X}$, then does $\mathbb{Z}$ have the expected resolution? In particular, is it true that $\mathbb{Z}$ is level if it is numerically possible?*

Furthermore, in line with the result of Lauze, on any (or at least "general") integral ACM variety $\mathbb{X}$, if $\mathbb{Z}$ is a general set of points on $\mathbb{X}$, then if it is numerically possible, does it follow that $\mathbb{Z}$ is level? And if it is numerically possible, does it follow that the additional generators of $\mathbb{Z}$ (those that are not generators of $\mathbb{X}$) are the expected number?

Part 2

Appendix: A Classification of Codimension Three Level Algebras of Low Socle Degree

APPENDIX A

Introduction and Notation

In this sequence of appendices, we record our investigations into what can be the h-vector of a codimension 3 level algebra. We actually are interested in recording a bit more – more precisely:

a) What can be the h-vector of a codimension 3 Artinian level algebra?
b) What can be the h-vector of a level set of points in $\mathbb{P}^3$?
c) What can be the h-vector of a codimension 3 Artinian level algebra with the Weak Lefschetz Property (WLP)?

Obviously any h-vector which satisfies b) satisfies a) – and we wonder if the answers to a), b) and c) are the same.

In this appendix we will consider all three questions for socle degree ≤ 4 (every type) and also for socle degree 5, type 2, and socle degree 6, type 2. In the remaining cases, i.e. socle degree 5 and every type other than 2, we consider only questions a) and c).

- For socle degree ≤ 4 (any type), socle degree 5 (type 2) and socle degree 6 (type 2) we find that a), b) and c) have the same answers.
- For socle degree 5 (any type $\neq 2$) we find that a) and c) have the same answer *except for one case* (which we cannot decide): see Table 5.7, item 30], where we find an example for a) but cannot verify c) for this h-vector.

The appendix is organized by following the same procedure for each socle degree and type. I.e. when we consider socle degree α and type β we first list (in **Table $\alpha.\beta$A**) all the h-vectors of Artinian algebras having socle degree α and last entry β (and which also satisfy the addition simple condition of Theorem A, ii) of the paper). We are extremely grateful to G. Dalzotto for the CoCoA programme which produces such lists.

With this given list we proceed to use all known theorems and ad-hoc arguments to eliminate h-vectors in this first table as the h-vector of a level algebra. All those h-vectors which remain are then recollected in another list, which is always labeled **Table $\alpha.\beta$**. We then proceed to show that all the vectors listed in **Table $\alpha.\beta$** exist for a) (and then for either b) or c) or both).

It follows then that the h-vectors listed in tables labeled **Table $\alpha.\beta$** are precisely the lists of h-vectors of level Artinian algebras of embedding dimension three, socle degree α and type β. We collect all such tables at the end of the appendix.

APPENDIX B

Socle Degree 6 and Type 2

Table 6.2A

1]	1, 3, 2, 2, 2, 2, 2	2]	1, 3, 3, 2, 2, 2, 2	3]	1, 3, 3, 3, 2, 2, 2
4]	1, 3, 3, 3, 3, 2, 2	5]	1, 3, 3, 3, 3, 3, 2	6]	1, 3, 3, 4, 2, 2, 2
7]	1, 3, 3, 4, 3, 2, 2	8]	1, 3, 3, 4, 3, 3, 2	9]	1, 3, 3, 4, 4, 2, 2
10]	1, 3, 3, 4, 4, 3, 2	11]	1, 3, 3, 4, 4, 4, 2	12]	1, 3, 3, 4, 5, 2, 2
13]	1, 3, 3, 4, 5, 3, 2	14]	1, 3, 3, 4, 5, 4, 2	15]	1, 3, 3, 4, 5, 5, 2
16]	1, 3, 3, 4, 5, 6, 2	17]	1, 3, 4, 2, 2, 2, 2	18]	1, 3, 4, 3, 2, 2, 2
19]	1, 3, 4, 3, 3, 2, 2	20]	1, 3, 4, 3, 3, 3, 2	21]	1, 3, 4, 4, 2, 2, 2
22]	1, 3, 4, 4, 3, 2, 2	23]	1, 3, 4, 4, 3, 3, 2	24]	1, 3, 4, 4, 4, 2, 2
25]	1, 3, 4, 4, 4, 3, 2	26]	1, 3, 4, 4, 4, 4, 2	27]	1, 3, 4, 4, 5, 2, 2
28]	1, 3, 4, 4, 5, 3, 2	29]	1, 3, 4, 4, 5, 4, 2	30]	1, 3, 4, 4, 5, 5, 2
31]	1, 3, 4, 4, 5, 6, 2	32]	1, 3, 4, 5, 2, 2, 2	33]	1, 3, 4, 5, 3, 2, 2
34]	1, 3, 4, 5, 3, 3, 2	35]	1, 3, 4, 5, 4, 2, 2	36]	1, 3, 4, 5, 4, 3, 2
37]	1, 3, 4, 5, 4, 4, 2	38]	1, 3, 4, 5, 5, 2, 2	39]	1, 3, 4, 5, 5, 3, 2
40]	1, 3, 4, 5, 5, 4, 2	41]	1, 3, 4, 5, 5, 5, 2	42]	1, 3, 4, 5, 5, 6, 2
43]	1, 3, 4, 5, 6, 2, 2	44]	1, 3, 4, 5, 6, 3, 2	45]	1, 3, 4, 5, 6, 4, 2
46]	1, 3, 4, 5, 6, 5, 2	47]	1, 3, 4, 5, 6, 6, 2	48]	1, 3, 5, 2, 2, 2, 2
49]	1, 3, 5, 3, 2, 2, 2	50]	1, 3, 5, 3, 3, 2, 2	51]	1, 3, 5, 3, 3, 3, 2
52]	1, 3, 5, 4, 2, 2, 2	53]	1, 3, 5, 4, 3, 2, 2	54]	1, 3, 5, 4, 3, 3, 2
55]	1, 3, 5, 4, 4, 2, 2	56]	1, 3, 5, 4, 4, 3, 2	57]	1, 3, 5, 4, 4, 4, 2
58]	1, 3, 5, 4, 5, 2, 2	59]	1, 3, 5, 4, 5, 3, 2	60]	1, 3, 5, 4, 5, 4, 2
61]	1, 3, 5, 4, 5, 5, 2	62]	1, 3, 5, 4, 5, 6, 2	63]	1, 3, 5, 5, 2, 2, 2
64]	1, 3, 5, 5, 3, 2, 2	65]	1, 3, 5, 5, 3, 3, 2	66]	1, 3, 5, 5, 4, 2, 2
67]	1, 3, 5, 5, 4, 3, 2	68]	1, 3, 5, 5, 4, 4, 2	69]	1, 3, 5, 5, 5, 2, 2
70]	1, 3, 5, 5, 5, 3, 2	71]	1, 3, 5, 5, 5, 4, 2	72]	1, 3, 5, 5, 5, 5, 2
73]	1, 3, 5, 5, 5, 6, 2	74]	1, 3, 5, 5, 6, 2, 2	75]	1, 3, 5, 5, 6, 3, 2
76]	1, 3, 5, 5, 6, 4, 2	77]	1, 3, 5, 5, 6, 5, 2	78]	1, 3, 5, 5, 6, 6, 2
79]	1, 3, 5, 6, 2, 2, 2	80]	1, 3, 5, 6, 3, 2, 2	81]	1, 3, 5, 6, 3, 3, 2
82]	1, 3, 5, 6, 4, 2, 2	83]	1, 3, 5, 6, 4, 3, 2	84]	1, 3, 5, 6, 4, 4, 2
85]	1, 3, 5, 6, 5, 2, 2	86]	1, 3, 5, 6, 5, 3, 2	87]	1, 3, 5, 6, 5, 4, 2
88]	1, 3, 5, 6, 5, 5, 2	89]	1, 3, 5, 6, 5, 6, 2	90]	1, 3, 5, 6, 6, 2, 2
91]	1, 3, 5, 6, 6, 3, 2	92]	1, 3, 5, 6, 6, 4, 2	93]	1, 3, 5, 6, 6, 5, 2
94]	1, 3, 5, 6, 6, 6, 2	95]	1, 3, 5, 6, 7, 2, 2	96]	1, 3, 5, 6, 7, 3, 2
97]	1, 3, 5, 6, 7, 4, 2	98]	1, 3, 5, 6, 7, 5, 2	99]	1, 3, 5, 6, 7, 6, 2
100]	1, 3, 5, 7, 2, 2, 2	101]	1, 3, 5, 7, 3, 2, 2	102]	1, 3, 5, 7, 3, 3, 2
103]	1, 3, 5, 7, 4, 2, 2	104]	1, 3, 5, 7, 4, 3, 2	105]	1, 3, 5, 7, 4, 4, 2
106]	1, 3, 5, 7, 5, 2, 2	107]	1, 3, 5, 7, 5, 3, 2	108]	1, 3, 5, 7, 5, 4, 2
109]	1, 3, 5, 7, 5, 5, 2	110]	1, 3, 5, 7, 5, 6, 2	111]	1, 3, 5, 7, 6, 2, 2

112]	1, 3, 5, 7, 6, 3, 2	113]	1, 3, 5, 7, 6, 4, 2	114]	1, 3, 5, 7, 6, 5, 2
115]	1, 3, 5, 7, 6, 6, 2	116]	1, 3, 5, 7, 7, 2, 2	117]	1, 3, 5, 7, 7, 3, 2
118]	1, 3, 5, 7, 7, 4, 2	119]	1, 3, 5, 7, 7, 5, 2	120]	1, 3, 5, 7, 7, 6, 2
121]	1, 3, 5, 7, 8, 2, 2	122]	1, 3, 5, 7, 8, 3, 2	123]	1, 3, 5, 7, 8, 4, 2
124]	1, 3, 5, 7, 8, 5, 2	125]	1, 3, 5, 7, 8, 6, 2	126]	1, 3, 5, 7, 9, 2, 2
127]	1, 3, 5, 7, 9, 3, 2	128]	1, 3, 5, 7, 9, 4, 2	129]	1, 3, 5, 7, 9, 5, 2
130]	1, 3, 5, 7, 9, 6, 2	131]	1, 3, 6, 2, 2, 2, 2	132]	1, 3, 6, 3, 2, 2, 2
133]	1, 3, 6, 3, 3, 2, 2	134]	1, 3, 6, 3, 3, 3, 2	135]	1, 3, 6, 4, 2, 2, 2
136]	1, 3, 6, 4, 3, 2, 2	137]	1, 3, 6, 4, 3, 3, 2	138]	1, 3, 6, 4, 4, 2, 2
139]	1, 3, 6, 4, 4, 3, 2	140]	1, 3, 6, 4, 4, 4, 2	141]	1, 3, 6, 4, 5, 2, 2
142]	1, 3, 6, 4, 5, 3, 2	143]	1, 3, 6, 4, 5, 4, 2	144]	1, 3, 6, 4, 5, 5, 2
145]	1, 3, 6, 4, 5, 6, 2	146]	1, 3, 6, 5, 2, 2, 2	147]	1, 3, 6, 5, 3, 2, 2
148]	1, 3, 6, 5, 3, 3, 2	149]	1, 3, 6, 5, 4, 2, 2	150]	1, 3, 6, 5, 4, 3, 2
151]	1, 3, 6, 5, 4, 4, 2	152]	1, 3, 6, 5, 5, 2, 2	153]	1, 3, 6, 5, 5, 3, 2
154]	1, 3, 6, 5, 5, 4, 2	155]	1, 3, 6, 5, 5, 5, 2	156]	1, 3, 6, 5, 5, 6, 2
157]	1, 3, 6, 5, 6, 2, 2	158]	1, 3, 6, 5, 6, 3, 2	159]	1, 3, 6, 5, 6, 4, 2
160]	1, 3, 6, 5, 6, 5, 2	161]	1, 3, 6, 5, 6, 6, 2	162]	1, 3, 6, 6, 2, 2, 2
163]	1, 3, 6, 6, 3, 2, 2	164]	1, 3, 6, 6, 3, 3, 2	165]	1, 3, 6, 6, 4, 2, 2
166]	1, 3, 6, 6, 4, 3, 2	167]	1, 3, 6, 6, 4, 4, 2	168]	1, 3, 6, 6, 5, 2, 2
169]	1, 3, 6, 6, 5, 3, 2	170]	1, 3, 6, 6, 5, 4, 2	171]	1, 3, 6, 6, 5, 5, 2
172]	1, 3, 6, 6, 5, 6, 2	173]	1, 3, 6, 6, 6, 2, 2	174]	1, 3, 6, 6, 6, 3, 2
175]	1, 3, 6, 6, 6, 4, 2	176]	1, 3, 6, 6, 6, 5, 2	177]	1, 3, 6, 6, 6, 6, 2
178]	1, 3, 6, 6, 7, 2, 2	179]	1, 3, 6, 6, 7, 3, 2	180]	1, 3, 6, 6, 7, 4, 2
181]	1, 3, 6, 6, 7, 5, 2	182]	1, 3, 6, 6, 7, 6, 2	183]	1, 3, 6, 7, 2, 2, 2
184]	1, 3, 6, 7, 3, 2, 2	185]	1, 3, 6, 7, 3, 3, 2	186]	1, 3, 6, 7, 4, 2, 2
187]	1, 3, 6, 7, 4, 3, 2	188]	1, 3, 6, 7, 4, 4, 2	189]	1, 3, 6, 7, 5, 2, 2
190]	1, 3, 6, 7, 5, 3, 2	191]	1, 3, 6, 7, 5, 4, 2	192]	1, 3, 6, 7, 5, 5, 2
193]	1, 3, 6, 7, 5, 6, 2	194]	1, 3, 6, 7, 6, 2, 2	195]	1, 3, 6, 7, 6, 3, 2
196]	1, 3, 6, 7, 6, 4, 2	197]	1, 3, 6, 7, 6, 5, 2	198]	1, 3, 6, 7, 6, 6, 2
199]	1, 3, 6, 7, 7, 2, 2	200]	1, 3, 6, 7, 7, 3, 2	201]	1, 3, 6, 7, 7, 4, 2
202]	1, 3, 6, 7, 7, 5, 2	203]	1, 3, 6, 7, 7, 6, 2	204]	1, 3, 6, 7, 8, 2, 2
205]	1, 3, 6, 7, 8, 3, 2	206]	1, 3, 6, 7, 8, 4, 2	207]	1, 3, 6, 7, 8, 5, 2
208]	1, 3, 6, 7, 8, 6, 2	209]	1, 3, 6, 7, 9, 2, 2	210]	1, 3, 6, 7, 9, 3, 2
211]	1, 3, 6, 7, 9, 4, 2	212]	1, 3, 6, 7, 9, 5, 2	213]	1, 3, 6, 7, 9, 6, 2
214]	1, 3, 6, 8, 2, 2, 2	215]	1, 3, 6, 8, 3, 2, 2	216]	1, 3, 6, 8, 3, 3, 2
217]	1, 3, 6, 8, 4, 2, 2	218]	1, 3, 6, 8, 4, 3, 2	219]	1, 3, 6, 8, 4, 4, 2
220]	1, 3, 6, 8, 5, 2, 2	221]	1, 3, 6, 8, 5, 3, 2	222]	1, 3, 6, 8, 5, 4, 2
223]	1, 3, 6, 8, 5, 5, 2	224]	1, 3, 6, 8, 5, 6, 2	225]	1, 3, 6, 8, 6, 2, 2
226]	1, 3, 6, 8, 6, 3, 2	227]	1, 3, 6, 8, 6, 4, 2	228]	1, 3, 6, 8, 6, 5, 2
229]	1, 3, 6, 8, 6, 6, 2	230]	1, 3, 6, 8, 7, 2, 2	231]	1, 3, 6, 8, 7, 3, 2
232]	1, 3, 6, 8, 7, 4, 2	233]	1, 3, 6, 8, 7, 5, 2	234]	1, 3, 6, 8, 7, 6, 2
235]	1, 3, 6, 8, 8, 2, 2	236]	1, 3, 6, 8, 8, 3, 2	237]	1, 3, 6, 8, 8, 4, 2
238]	1, 3, 6, 8, 8, 5, 2	239]	1, 3, 6, 8, 8, 6, 2	240]	1, 3, 6, 8, 9, 2, 2
241]	1, 3, 6, 8, 9, 3, 2	242]	1, 3, 6, 8, 9, 4, 2	243]	1, 3, 6, 8, 9, 5, 2
244]	1, 3, 6, 8, 9, 6, 2	245]	1, 3, 6, 8, 10, 2, 2	246]	1, 3, 6, 8, 10, 3, 2
247]	1, 3, 6, 8, 10, 4, 2	248]	1, 3, 6, 8, 10, 5, 2	249]	1, 3, 6, 8, 10, 6, 2
250]	1, 3, 6, 9, 2, 2, 2	251]	1, 3, 6, 9, 3, 2, 2	252]	1, 3, 6, 9, 3, 3, 2

B. SOCLE DEGREE 6 AND TYPE 2

253]	1, 3, 6, 9, 4, 2, 2	254]	1, 3, 6, 9, 4, 3, 2	255]	1, 3, 6, 9, 4, 4, 2
256]	1, 3, 6, 9, 5, 2, 2	257]	1, 3, 6, 9, 5, 3, 2	258]	1, 3, 6, 9, 5, 4, 2
259]	1, 3, 6, 9, 5, 5, 2	260]	1, 3, 6, 9, 5, 6, 2	261]	1, 3, 6, 9, 6, 2, 2
262]	1, 3, 6, 9, 6, 3, 2	263]	1, 3, 6, 9, 6, 4, 2	264]	1, 3, 6, 9, 6, 5, 2
265]	1, 3, 6, 9, 6, 6, 2	266]	1, 3, 6, 9, 7, 2, 2	267]	1, 3, 6, 9, 7, 3, 2
268]	1, 3, 6, 9, 7, 4, 2	269]	1, 3, 6, 9, 7, 5, 2	270]	1, 3, 6, 9, 7, 6, 2
271]	1, 3, 6, 9, 8, 2, 2	272]	1, 3, 6, 9, 8, 3, 2	273]	1, 3, 6, 9, 8, 4, 2
274]	1, 3, 6, 9, 8, 5, 2	275]	1, 3, 6, 9, 8, 6, 2	276]	1, 3, 6, 9, 9, 2, 2
277]	1, 3, 6, 9, 9, 3, 2	278]	1, 3, 6, 9, 9, 4, 2	279]	1, 3, 6, 9, 9, 5, 2
280]	1, 3, 6, 9, 9, 6, 2	281]	1, 3, 6, 9, 10, 2, 2	282]	1, 3, 6, 9, 10, 3, 2
283]	1, 3, 6, 9, 10, 4, 2	284]	1, 3, 6, 9, 10, 5, 2	285]	1, 3, 6, 9, 10, 6, 2
286]	1, 3, 6, 9, 11, 2, 2	287]	1, 3, 6, 9, 11, 3, 2	288]	1, 3, 6, 9, 11, 4, 2
289]	1, 3, 6, 9, 11, 5, 2	290]	1, 3, 6, 9, 11, 6, 2	291]	1, 3, 6, 9, 12, 2, 2
292]	1, 3, 6, 9, 12, 3, 2	293]	1, 3, 6, 9, 12, 4, 2	294]	1, 3, 6, 9, 12, 5, 2
295]	1, 3, 6, 9, 12, 6, 2	296]	1, 3, 6, 10, 2, 2, 2	297]	1, 3, 6, 10, 3, 2, 2
298]	1, 3, 6, 10, 3, 3, 2	299]	1, 3, 6, 10, 4, 2, 2	300]	1, 3, 6, 10, 4, 3, 2
301]	1, 3, 6, 10, 4, 4, 2	302]	1, 3, 6, 10, 5, 2, 2	303]	1, 3, 6, 10, 5, 3, 2
304]	1, 3, 6, 10, 5, 4, 2	305]	1, 3, 6, 10, 5, 5, 2	306]	1, 3, 6, 10, 5, 6, 2
307]	1, 3, 6, 10, 6, 2, 2	308]	1, 3, 6, 10, 6, 3, 2	309]	1, 3, 6, 10, 6, 4, 2
310]	1, 3, 6, 10, 6, 5, 2	311]	1, 3, 6, 10, 6, 6, 2	312]	1, 3, 6, 10, 7, 2, 2
313]	1, 3, 6, 10, 7, 3, 2	314]	1, 3, 6, 10, 7, 4, 2	315]	1, 3, 6, 10, 7, 5, 2
316]	1, 3, 6, 10, 7, 6, 2	317]	1, 3, 6, 10, 8, 2, 2	318]	1, 3, 6, 10, 8, 3, 2
319]	1, 3, 6, 10, 8, 4, 2	320]	1, 3, 6, 10, 8, 5, 2	321]	1, 3, 6, 10, 8, 6, 2
322]	1, 3, 6, 10, 9, 2, 2	323]	1, 3, 6, 10, 9, 3, 2	324]	1, 3, 6, 10, 9, 4, 2
325]	1, 3, 6, 10, 9, 5, 2	326]	1, 3, 6, 10, 9, 6, 2	327]	1, 3, 6, 10, 10, 2, 2
328]	1, 3, 6, 10, 10, 3, 2	329]	1, 3, 6, 10, 10, 4, 2	330]	1, 3, 6, 10, 10, 5, 2
331]	1, 3, 6, 10, 10, 6, 2	332]	1, 3, 6, 10, 11, 2, 2	333]	1, 3, 6, 10, 11, 3, 2
334]	1, 3, 6, 10, 11, 4, 2	335]	1, 3, 6, 10, 11, 5, 2	336]	1, 3, 6, 10, 11, 6, 2
337]	1, 3, 6, 10, 12, 2, 2	338]	1, 3, 6, 10, 12, 3, 2	339]	1, 3, 6, 10, 12, 4, 2
340]	1, 3, 6, 10, 12, 5, 2	341]	1, 3, 6, 10, 12, 6, 2		

The reader should note that we have already eliminated some O-sequences. From our interpretation in terms of inverse systems we know that if we have a level algebra of type 2 in socle degree s then a level h-vector $(1, 3, \ldots, h_{s-2}, h_{s-1}, 2)$ must have $h_{s-1} \leq 6$ and $h_{s-2} \leq 12$, so we immediately eliminated all 7-tuples that didn't have that property. The resulting list, as one can see, has 341 possibilities.

Non-existence

Eliminated by Remark 2.12, b). $(\mathbf{1}, \mathbf{3}, \ldots, \mathbf{2}, \mathbf{2})$: 1], 2], 3], 4], 6], 7], 9], 12], 17], 18], 19], 21], 22], 24], 27], 32], 33], 35], 38], 43], 48], 49], 50], 52], 53], 55], 58], 63], 64], 66], 69], 74], 79], 80], 82], 85], 90], 95], 100], 101], 103], 106], 111], 116], 121], 126], 131], 132], 133], 135], 136], 138], 141], 146], 147], 149], 152], 157], 162], 163], 165], 168], 173], 178], 183], 184], 186], 189], 194], 199], 204], 209], 214], 215], 217], 220], 225], 230], 235], 240], 245], 250], 251], 253], 256], 261], 266], 271], 276], 281], 286], 291], 296], 297], 299], 302], 307], 312], 317], 322], 327], 332], 337].

$(\mathbf{1}, \mathbf{3}, \ldots, \geq \mathbf{8}, \mathbf{4}, \mathbf{2})$: 123], 128], 206], 211], 237], 242], 247], 273], 278], 283], 288], 293], 319], 324], 329], 334], 339].

$(1, 3, \ldots, \geq 10, 5, 2)$: 248], 284], 289], 294], 330], 335], 340].

$(1, 3, 6, \ldots, 4, 3, 2)$: 150], 166], 187], 218], 254], 300].

$(1, 3, 6, 5, 5, 5, 2)$: 155].

Eliminated by Example 2.13. $(1, 3, \ldots, 5, 3, 2)$: 13], 28], 39], 59], 70], 86], 107], 142], 153], 169], 190], 221], 257], 303].

Eliminated by Example 3.1. $(1, 3, 4, 5, 4, 4, 2)$: 37].

Eliminated by Corollary 3.5. 8], 10], 11], 14], 15], 16], 23], 29], 30], 31], 34], 42], 54], 60], 61], 62], 65], 73], 76], 77], 78], 81], 89], 102], 110], 137], 143], 144], 145], 148], 156], 159], 160], 161], 164], 172], 185], 193], 212], 213], 216], 224], 252], 260], 298], 306].

Eliminated by Proposition 3.8.
Part b): for $d = 3$: 20], 51], 134].
Part b): for $d = 4$: 68], 84], 105], 151], 167], 188], 219], 255], 301].
Part c): for $d = 3$: 56], 57], 139], 140].
Part c): for $d = 4$: 88], 109], 171], 192], 223], 259], 305].

Eliminated by Remark 3.10. $(1, 3, \ldots, \geq 6, 3, 2)$: 44], 75], 91], 96], 112], 117], 122], 127], 158], 174], 179], 195], 200], 205], 210], 226], 231], 236], 241], 246], 262], 267], 272], 277], 282], 287], 292], 308], 313], 318], 323], 328], 333], 338].

Eliminated by Example 3.11. 108], 115], 198], 228], 270], 309], 316].

Eliminated by Example 3.12. 129], 263], 315].

Eliminated by Example 5.7. $(1, 3, 5, \ldots, 4, 3, 2)$: 67], 83], 104].
$(1, 3, 5, \ldots, 7, 4, 2)$: 97], 118].
$(1, 3, 6, \ldots, 5, 4, 2)$: 154], 170], 191], 222], 258], 304].

Eliminated by Example 5.9. $(1, 3, 4, 5, \ldots, 6, 4, 2)$: 45].

The following are eliminated by "non-cancelation", but are not covered by any of our theorems.

180]	1, 3, 6, 6, 7, 4, 2	(socle in degree 2)
181]	1, 3, 6, 6, 7, 5, 2	(socle in degree 2)
182]	1, 3, 6, 6, 7, 6, 2	(socle in degree 2)
229]	1, 3, 6, 8, 6, 6, 2	(socle in degree 3)
264]	1, 3, 6, 9, 6, 5, 2	(socle in degree 3)
265]	1, 3, 6, 9, 6, 6, 2	(socle in degree 3)
310]	1, 3, 6, 10, 6, 5, 2	(socle in degree 3)
311]	1, 3, 6, 10, 6, 6, 2	(socle in degree 3)

We now show that 47] does not exist either. Our proof is similar to that given in [**14**], Proposition 2.7.

Proof. Suppose there were a level algebra A whose h-vector was:
$$h = \mathbf{H}(A) = (1, 3, 4, 5, 6, 6, 2).$$

We will write $A = R/I$ where $R = k[x, y, z]$ and let $S = k[X, Y, Z]$.

From the h-vector we see that the 2-dimensional space, I_2, has to be generated by two quadratic forms which (because of h_3) must share a common factor. There are two cases to consider:

Case 1. $I_2 = \langle xy, xz \rangle$.

Obviously $I_6 \supseteq R_4 I_2$ and so $(I^{-1})_6 \subseteq (R_4 I_2)^\perp$. Notice that $(R_4 I_2)^\perp$ is generated by all the monomials of degree 6 in Y and Z plus X^6.

Let $F, G \in S_6$ generate I^{-1} then $\langle F, G \rangle \subseteq (R_4 I_2)^\perp$. With no loss of generality we can assume that only F involves X^6. But then G is a polynomial in only two variables and so can have, at most, 2 linearly independent first derivatives. Since F can have at most 3 linearly independent first derivatives we see that F and G can generate, at most, a subspace of S_5 of dimension at most 5. But, $h_5 = 6$ and so this is impossible.

Case 2. $I_2 = \langle x^2, xy \rangle$.

This is very similar to Case 1. This time we note that $(R_4 I_2)^\perp$ is generated by all the monomials in Y and Z of degree 6, plus XZ^5. As in Case 1, we can assume that at most F involves the monomial XZ^5 and thus G is a polynomial in only two variables. The proof goes as in Case 1 and so we obtain a contradiction also in this Case.

Thus, such a level algebra does not exist. □

Existence

We now turn to the existence question. We show that the remaining 58 h-vectors are all the h-vectors of an Artinian level algebra of socle degree 6 and type 2. Those 58 h-vectors are:

Table 6.2

5]	1, 3, 3, 3, 3, 3, 2	25]	1, 3, 4, 4, 4, 3, 2	26]	1, 3, 4, 4, 4, 4, 2
36]	1, 3, 4, 5, 4, 3, 2	40]	1, 3, 4, 5, 5, 4, 2	41]	1, 3, 4, 5, 5, 5, 2
46]	1, 3, 4, 5, 6, 5, 2	71]	1, 3, 5, 5, 5, 4, 2	72]	1, 3, 5, 5, 5, 5, 2
87]	1, 3, 5, 6, 5, 4, 2	92]	1, 3, 5, 6, 6, 4, 2	93]	1, 3, 5, 6, 6, 5, 2
94]	1, 3, 5, 6, 6, 6, 2	98]	1, 3, 5, 6, 7, 5, 2	99]	1, 3, 5, 6, 7, 6, 2
113]	1, 3, 5, 7, 6, 4, 2	114]	1, 3, 5, 7, 6, 5, 2	119]	1, 3, 5, 7, 7, 5, 2
120]	1, 3, 5, 7, 7, 6, 2	124]	1, 3, 5, 7, 8, 5, 2	125]	1, 3, 5, 7, 8, 6, 2
130]	1, 3, 5, 7, 9, 6, 2	175]	1, 3, 6, 6, 6, 4, 2	176]	1, 3, 6, 6, 6, 5, 2
177]	1, 3, 6, 6, 6, 6, 2	196]	1, 3, 6, 7, 6, 4, 2	197]	1, 3, 6, 7, 6, 5, 2
201]	1, 3, 6, 7, 7, 4, 2	202]	1, 3, 6, 7, 7, 5, 2	203]	1, 3, 6, 7, 7, 6, 2
207]	1, 3, 6, 7, 8, 5, 2	208]	1, 3, 6, 7, 8, 6, 2	227]	1, 3, 6, 8, 6, 4, 2
232]	1, 3, 6, 8, 7, 4, 2	233]	1, 3, 6, 8, 7, 5, 2	234]	1, 3, 6, 8, 7, 6, 2
238]	1, 3, 6, 8, 8, 5, 2	239]	1, 3, 6, 8, 8, 6, 2	243]	1, 3, 6, 8, 9, 5, 2
244]	1, 3, 6, 8, 9, 6, 2	249]	1, 3, 6, 8, 10, 6, 2	268]	1, 3, 6, 9, 7, 4, 2
269]	1, 3, 6, 9, 7, 5, 2	274]	1, 3, 6, 9, 8, 5, 2	275]	1, 3, 6, 9, 8, 6, 2
279]	1, 3, 6, 9, 9, 5, 2	280]	1, 3, 6, 9, 9, 6, 2	285]	1, 3, 6, 9, 10, 6, 2
290]	1, 3, 6, 9, 11, 6, 2	295]	1, 3, 6, 9, 12, 6, 2	314]	1, 3, 6, 10, 7, 4, 2
320]	1, 3, 6, 10, 8, 5, 2	321]	1, 3, 6, 10, 8, 6, 2	325]	1, 3, 6, 10, 9, 5, 2
326]	1, 3, 6, 10, 9, 6, 2	331]	1, 3, 6, 10, 10, 6, 2	336]	1, 3, 6, 10, 11, 6, 2
341]	1, 3, 6, 10, 12, 6, 2				

The Linked-Sum Construction. This was described in subsection 5.4. For each example below we start with a level set of points (using **Remark 5.26**), $\mathbb{Z}$, in $\mathbb{P}^2$ (actually in $\mathbb{A}^2 \subset \mathbb{P}^2$) which we partition into two subsets $\mathbb{X}$ and $\mathbb{Y}$. The ring $A = k[x_0, x_1, x_2]/(I_{\mathbb{X}} + I_{\mathbb{Y}})$ is the level algebra with the desired h-vector. The points of $\mathbb{X}$ will be denoted with •'s and those of $\mathbb{Y}$ with ∗'s. Next to each diagram we give the Hilbert functions of all the relevant rings.

Using the "linked-sum" construction we construct an Artinian k-algebra for EVERY possible h-vector in Table 2 above.

5]. $\mathbb{Z} = \left\{\begin{matrix} * & * & • & • & \\ * & * & * & • & * \\ * & * & * & * & * \\ * & * & * & * & * \\ * & * & * & * & * \end{matrix}\right\}$

$\mathbf{H}(\mathbb{Z}, -)$: 1 3 6 10 15 19 22 24 →
$\mathbf{H}(\mathbb{X}, -)$: 1 3 3 3 3 3 3 3 →
$\mathbf{H}(\mathbb{Y}, -)$: 1 3 6 10 15 19 21 21 →
$\mathbf{H}(A, -)$: 1 3 3 3 3 3 2 0 →.

25]. $\mathbb{Z} = \left\{\begin{matrix} * & • & • & • & & \\ * & * & * & • & & \\ * & * & * & * & * & * \\ * & * & * & * & * & * \\ * & * & * & * & * & * \end{matrix}\right\}$

$\mathbf{H}(\mathbb{Z}, -)$: 1 3 6 10 15 20 24 26 →
$\mathbf{H}(\mathbb{X}, -)$: 1 3 4 4 4 4 4 4 →
$\mathbf{H}(\mathbb{Y}, -)$: 1 3 6 10 15 19 22 22 →
$\mathbf{H}(A, -)$: 1 3 4 4 4 3 2 0 →.

26]. $\mathbb{Z} = \left\{\begin{matrix} • & • & * & * & \\ • & • & * & * & * \\ * & * & * & * & * \\ * & * & * & * & * \\ * & * & * & * & * \end{matrix}\right\}$

$\mathbf{H}(\mathbb{Z}, -)$: 1 3 6 10 15 19 22 24 →
$\mathbf{H}(\mathbb{X}, -)$: 1 3 4 4 4 4 4 4 →
$\mathbf{H}(\mathbb{Y}, -)$: 1 3 6 10 15 19 20 20 →
$\mathbf{H}(A, -)$: 1 3 4 4 4 4 2 0 →.

36]. $\mathbb{Z} = \left\{\begin{matrix} • & • & • & • & & \\ • & * & * & * & & \\ * & * & * & * & * & * \\ * & * & * & * & * & * \\ * & * & * & * & * & * \end{matrix}\right\}$

$\mathbf{H}(\mathbb{Z}, -)$: 1 3 6 10 15 19 22 24 →
$\mathbf{H}(\mathbb{X}, -)$: 1 3 4 5 5 5 5 5 →
$\mathbf{H}(\mathbb{Y}, -)$: 1 3 6 10 14 17 19 19 →
$\mathbf{H}(A, -)$: 1 3 4 5 4 3 2 0 →.

40]. $\mathbb{Z} = \left\{\begin{matrix} * & * & * & • & & \\ * & * & * & • & & \\ * & * & * & • & • & • \\ * & * & * & * & * & * \\ * & * & * & * & * & * \end{matrix}\right\}$

$\mathbf{H}(\mathbb{Z}, -)$: 1 3 6 10 15 20 24 26 →
$\mathbf{H}(\mathbb{X}, -)$: 1 3 4 5 5 5 5 5 →
$\mathbf{H}(\mathbb{Y}, -)$: 1 3 6 10 15 19 21 21 →
$\mathbf{H}(A, -)$: 1 3 4 5 5 4 2 0 →.

41]. $\mathbb{Z} = \left\{\begin{matrix} * & * & * & * & & \\ * & * & * & • & & \\ * & * & • & • & • & • \\ * & * & * & * & * & * \\ * & * & * & * & * & * \end{matrix}\right\}$

$\mathbf{H}(\mathbb{Z}, -)$: 1 3 6 10 15 20 24 26 →
$\mathbf{H}(\mathbb{X}, -)$: 1 3 4 5 5 5 5 5 →
$\mathbf{H}(\mathbb{Y}, -)$: 1 3 6 10 15 20 21 21 →
$\mathbf{H}(A, -)$: 1 3 4 5 5 5 2 0 →.

46]. $\mathbb{Z} = \left\{\begin{matrix} • & • & * & * & & \\ • & * & * & * & & \\ • & * & * & * & * & * \\ • & * & * & * & * & * \\ • & * & * & * & * & * \end{matrix}\right\}$

$\mathbf{H}(\mathbb{Z}, -)$: 1 3 6 10 15 20 24 26 →
$\mathbf{H}(\mathbb{X}, -)$: 1 3 4 5 6 6 6 6 →
$\mathbf{H}(\mathbb{Y}, -)$: 1 3 6 10 15 19 20 20 →
$\mathbf{H}(A, -)$: 1 3 4 5 6 5 2 0 →.

B. SOCLE DEGREE 6 AND TYPE 2

71].
$\mathbb{Z} = \left\{ \begin{matrix} * & * & * & \bullet & & \\ * & * & * & \bullet & & \\ * & * & * & \bullet & \bullet & \bullet \\ * & * & * & * & * & * \\ * & * & * & * & * & * \end{matrix} \right\}$

$\mathbf{H}(\mathbb{Z},-)$: 1 3 6 10 15 20 24 26 →
$\mathbf{H}(\mathbb{X},-)$: 1 3 5 5 5 5 5 5 →
$\mathbf{H}(\mathbb{Y},-)$: 1 3 6 10 15 19 21 21 →
$\mathbf{H}(A,-)$: 1 3 5 5 5 4 2 0 →.

72].
$\mathbb{Z} = \left\{ \begin{matrix} * & * & * & * & & \\ * & * & * & * & \bullet & \bullet \\ * & * & \bullet & * & \bullet & \bullet \\ * & * & * & * & * & \\ * & * & * & * & * & \end{matrix} \right\}$

$\mathbf{H}(\mathbb{Z},-)$: 1 3 6 10 15 19 22 24 →
$\mathbf{H}(\mathbb{X},-)$: 1 3 5 5 5 5 5 5 →
$\mathbf{H}(\mathbb{Y},-)$: 1 3 6 10 15 19 19 19 →
$\mathbf{H}(A,-)$: 1 3 5 5 5 5 2 0 →.

87].
$\mathbb{Z} = \left\{ \begin{matrix} \bullet & \bullet & \bullet & \bullet & & \\ \bullet & * & * & * & & \\ \bullet & * & * & * & * & * \\ * & * & * & * & * & * \\ * & * & * & * & * & * \end{matrix} \right\}$

$\mathbf{H}(\mathbb{Z},-)$: 1 3 6 10 15 20 24 26 →
$\mathbf{H}(\mathbb{X},-)$: 1 3 5 6 6 6 6 6 →
$\mathbf{H}(\mathbb{Y},-)$: 1 3 6 10 14 18 20 20 →
$\mathbf{H}(A,-)$: 1 3 5 6 5 4 2 0 →.

92].
$\mathbb{Z} = \left\{ \begin{matrix} * & * & * & * & & \\ * & * & * & * & * & \\ * & * & * & \bullet & \bullet & \\ * & * & * & \bullet & \bullet & \\ * & * & * & \bullet & \bullet & \end{matrix} \right\}$

$\mathbf{H}(\mathbb{Z},-)$: 1 3 6 10 15 19 22 24 →
$\mathbf{H}(\mathbb{X},-)$: 1 3 5 6 6 6 6 6 →
$\mathbf{H}(\mathbb{Y},-)$: 1 3 6 10 15 17 18 18 →
$\mathbf{H}(A,-)$: 1 3 5 6 6 4 2 0 →.

93].
$\mathbb{Z} = \left\{ \begin{matrix} \bullet & \bullet & * & * & & \\ \bullet & \bullet & * & * & * & \\ \bullet & \bullet & * & * & * & \\ * & * & * & * & * & \\ * & * & * & * & * & \end{matrix} \right\}$

$\mathbf{H}(\mathbb{Z},-)$: 1 3 6 10 15 19 22 24 →
$\mathbf{H}(\mathbb{X},-)$: 1 3 5 6 6 6 6 6 →
$\mathbf{H}(\mathbb{Y},-)$: 1 3 6 10 15 18 18 18 →
$\mathbf{H}(A,-)$: 1 3 5 6 6 5 2 0 →.

94].
$\mathbb{Z} = \left\{ \begin{matrix} \bullet & * & * & * & & \\ \bullet & * & * & \bullet & & \\ \bullet & * & * & * & * & * \\ \bullet & * & \bullet & * & * & * \\ * & * & * & * & * & * \end{matrix} \right\}$

$\mathbf{H}(\mathbb{Z},-)$: 1 3 6 10 15 20 24 26 →
$\mathbf{H}(\mathbb{X},-)$: 1 3 5 6 6 6 6 6 →
$\mathbf{H}(\mathbb{Y},-)$: 1 3 6 10 15 20 20 20 →
$\mathbf{H}(A,-)$: 1 3 5 6 6 6 2 0 →.

98].
$\mathbb{Z} = \left\{ \begin{matrix} \bullet & \bullet & * & * & & \\ \bullet & \bullet & * & * & & \\ \bullet & * & * & * & * & * \\ \bullet & * & * & * & * & * \\ \bullet & * & * & * & * & * \end{matrix} \right\}$

$\mathbf{H}(\mathbb{Z},-)$: 1 3 6 10 15 20 24 26 →
$\mathbf{H}(\mathbb{X},-)$: 1 3 5 6 7 7 7 7 →
$\mathbf{H}(\mathbb{Y},-)$: 1 3 6 10 15 18 19 19 →
$\mathbf{H}(A,-)$: 1 3 5 6 7 5 2 0 →.

99].
$\mathbb{Z} = \left\{ \begin{matrix} \bullet & \bullet & * & * & & \\ * & \bullet & * & * & & \\ * & \bullet & * & * & * & * \\ * & \bullet & * & \bullet & * & * \\ * & \bullet & * & * & * & * \end{matrix} \right\}$

$\mathbf{H}(\mathbb{Z},-)$: 1 3 6 10 15 20 24 26 →
$\mathbf{H}(\mathbb{X},-)$: 1 3 5 6 7 7 7 7 →
$\mathbf{H}(\mathbb{Y},-)$: 1 3 6 10 15 19 19 19 →
$\mathbf{H}(A,-)$: 1 3 5 6 7 6 2 0 →.

113]. $\mathbb{Z} = \left\{\begin{matrix} \bullet & \bullet & \bullet & \bullet & \\ \bullet & \bullet & \bullet & * & * \\ * & * & * & * & * \\ * & * & * & * & * \\ * & * & * & * & * \end{matrix}\right\}$
$\mathbf{H}(\mathbb{Z},-)$: 1 3 6 10 15 19 22 24 $\to$
$\mathbf{H}(\mathbb{X},-)$: 1 3 5 7 7 7 7 7 $\to$
$\mathbf{H}(\mathbb{Y},-)$: 1 3 6 10 14 16 17 17 $\to$
$\mathbf{H}(A,-)$: 1 3 5 7 6 4 2 0 $\to .$

114]. $\mathbb{Z} = \left\{\begin{matrix} \bullet & \bullet & \bullet & \bullet & & \\ * & * & * & \bullet & & \\ * & * & * & \bullet & * & * \\ * & * & * & \bullet & * & * \\ * & * & * & * & * & * \end{matrix}\right\}$
$\mathbf{H}(\mathbb{Z},-)$: 1 3 6 10 15 20 24 26 $\to$
$\mathbf{H}(\mathbb{X},-)$: 1 3 5 7 7 7 7 7 $\to$
$\mathbf{H}(\mathbb{Y},-)$: 1 3 6 10 14 18 19 19 $\to$
$\mathbf{H}(A,-)$: 1 3 5 7 6 5 2 0 $\to .$

119]. $\mathbb{Z} = \left\{\begin{matrix} \bullet & \bullet & * & * & \\ \bullet & \bullet & * & * & * \\ \bullet & \bullet & * & * & * \\ \bullet & * & * & * & * \\ * & * & * & * & * \end{matrix}\right\}$
$\mathbf{H}(\mathbb{Z},-)$: 1 3 6 10 15 19 22 24 $\to$
$\mathbf{H}(\mathbb{X},-)$: 1 3 5 7 7 7 7 7 $\to$
$\mathbf{H}(\mathbb{Y},-)$: 1 3 6 10 15 17 17 17 $\to$
$\mathbf{H}(A,-)$: 1 3 5 7 7 5 2 0 $\to .$

120]. $\mathbb{Z} = \left\{\begin{matrix} \bullet & * & * & \bullet & & \\ \bullet & * & * & \bullet & & \\ * & * & * & \bullet & * & * \\ * & * & * & \bullet & * & * \\ \bullet & * & * & * & * & * \end{matrix}\right\}$
$\mathbf{H}(\mathbb{Z},-)$: 1 3 6 10 15 20 24 26 $\to$
$\mathbf{H}(\mathbb{X},-)$: 1 3 5 7 7 7 7 7 $\to$
$\mathbf{H}(\mathbb{Y},-)$: 1 3 6 10 15 19 19 19 $\to$
$\mathbf{H}(A,-)$: 1 3 5 7 7 6 2 0 $\to .$

124]. $\mathbb{Z} = \left\{\begin{matrix} * & \bullet & & & & \\ * & * & & & & \\ * & \bullet & & & & \\ * & * & * & * & * & \\ * & \bullet & * & * & * & \\ * & \bullet & \bullet & \bullet & \bullet & \\ * & \bullet & * & * & * & \end{matrix}\right\}$
$\mathbf{H}(\mathbb{Z},-)$: 1 3 6 10 15 20 24 26 $\to$
$\mathbf{H}(\mathbb{X},-)$: 1 3 5 7 8 8 8 8 $\to$
$\mathbf{H}(\mathbb{Y},-)$: 1 3 6 10 15 17 18 18 $\to$
$\mathbf{H}(A,-)$: 1 3 5 7 8 5 2 0 $\to .$

125]. $\mathbb{Z} = \left\{\begin{matrix} \bullet & \bullet & * & * & & \\ \bullet & \bullet & * & * & & \\ \bullet & \bullet & * & * & * & * \\ \bullet & * & * & * & * & * \\ \bullet & * & * & * & * & * \end{matrix}\right\}$
$\mathbf{H}(\mathbb{Z},-)$: 1 3 6 10 15 20 24 26 $\to$
$\mathbf{H}(\mathbb{X},-)$: 1 3 5 7 8 8 8 8 $\to$
$\mathbf{H}(\mathbb{Y},-)$: 1 3 6 10 15 18 18 18 $\to$
$\mathbf{H}(A,-)$: 1 3 5 7 8 6 2 0 $\to .$

130]. $\mathbb{Z} = \left\{\begin{matrix} * & * & \bullet & * & & \\ * & * & \bullet & * & & \\ * & * & \bullet & * & * & * \\ \bullet & \bullet & \bullet & * & \bullet & \bullet \\ * & * & \bullet & * & * & * \end{matrix}\right\}$
$\mathbf{H}(\mathbb{Z},-)$: 1 3 6 10 15 20 24 26 $\to$
$\mathbf{H}(\mathbb{X},-)$: 1 3 5 7 9 9 9 9 $\to$
$\mathbf{H}(\mathbb{Y},-)$: 1 3 6 10 15 17 17 17 $\to$
$\mathbf{H}(A,-)$: 1 3 5 7 9 6 2 0 $\to .$

175]. $\mathbb{Z} = \left\{\begin{matrix} \bullet & * & \bullet & \bullet & & \\ \bullet & * & * & \bullet & & \\ \bullet & * & * & * & * & * \\ * & * & * & * & * & * \\ * & * & * & * & * & * \end{matrix}\right\}$
$\mathbf{H}(\mathbb{Z},-)$: 1 3 6 10 15 20 24 26 $\to$
$\mathbf{H}(\mathbb{X},-)$: 1 3 6 6 6 6 6 6 $\to$
$\mathbf{H}(\mathbb{Y},-)$: 1 3 6 10 15 18 20 20 $\to$
$\mathbf{H}(A,-)$: 1 3 6 6 6 4 2 0 $\to .$

B. SOCLE DEGREE 6 AND TYPE 2

176].
$$\mathbb{Z} = \left\{ \begin{matrix} \bullet & * & * & * & \\ * & \bullet & * & * & \bullet \\ * & * & * & * & * \\ * & * & \bullet & * & * \\ * & \bullet & \bullet & * & * \end{matrix} \right\}$$

$\mathbf{H}(\mathbb{Z}, -)$: 1 3 6 10 15 19 22 24 $\to$
$\mathbf{H}(\mathbb{X}, -)$: 1 3 6 6 6 6 6 6 $\to$
$\mathbf{H}(\mathbb{Y}, -)$: 1 3 6 10 15 18 18 18 $\to$
$\mathbf{H}(A, -)$: 1 3 6 6 6 5 2 0 $\to$.

177].
$$\mathbb{Z} = \left\{ \begin{matrix} * & * & * & * & & \\ * & * & \bullet & * & & \\ * & * & * & * & * & * \\ \bullet & * & * & * & \bullet & * \\ * & * & \bullet & * & \bullet & \bullet \end{matrix} \right\}$$

$\mathbf{H}(\mathbb{Z}, -)$: 1 3 6 10 15 20 24 26 $\to$
$\mathbf{H}(\mathbb{X}, -)$: 1 3 6 6 6 6 6 6 $\to$
$\mathbf{H}(\mathbb{Y}, -)$: 1 3 6 10 15 20 20 20 $\to$
$\mathbf{H}(A, -)$: 1 3 6 6 6 6 2 0 $\to$.

196].
$$\mathbb{Z} = \left\{ \begin{matrix} \bullet & \bullet & * & \bullet & & \\ \bullet & * & \bullet & \bullet & & \\ * & * & * & * & * & * \\ * & * & * & * & * & * \\ \bullet & * & * & * & * & * \end{matrix} \right\}$$

$\mathbf{H}(\mathbb{Z}, -)$: 1 3 6 10 15 20 24 26 $\to$
$\mathbf{H}(\mathbb{X}, -)$: 1 3 6 7 7 7 7 7 $\to$
$\mathbf{H}(\mathbb{Y}, -)$: 1 3 6 10 14 17 19 19 $\to$
$\mathbf{H}(A, -)$: 1 3 6 7 6 4 2 0 $\to$.

197].
$$\mathbb{Z} = \left\{ \begin{matrix} * & * & \bullet & * & & \\ \bullet & \bullet & \bullet & \bullet & & \\ * & * & * & * & * & * \\ * & \bullet & * & * & \bullet & * \\ * & * & * & * & * & * \end{matrix} \right\}$$

$\mathbf{H}(\mathbb{Z}, -)$: 1 3 6 10 15 20 24 26 $\to$
$\mathbf{H}(\mathbb{X}, -)$: 1 3 6 7 7 7 7 7 $\to$
$\mathbf{H}(\mathbb{Y}, -)$: 1 3 6 10 14 18 19 19 $\to$
$\mathbf{H}(A, -)$: 1 3 6 7 6 5 2 0 $\to$.

201].
$$\mathbb{Z} = \left\{ \begin{matrix} * & & & \\ * & & & \\ * & & & \\ * & * & \bullet & \bullet \\ * & * & * & \bullet \\ * & * & * & * \\ * & * & * & * \\ \bullet & \bullet & \bullet & \bullet \end{matrix} \right\}$$

$\mathbf{H}(\mathbb{Z}, -)$: 1 3 6 10 14 18 21 23 $\to$
$\mathbf{H}(\mathbb{X}, -)$: 1 3 6 7 7 7 7 7 $\to$
$\mathbf{H}(\mathbb{Y}, -)$: 1 3 6 10 14 15 16 16 $\to$
$\mathbf{H}(A, -)$: 1 3 6 7 7 4 2 0 $\to$.

202].
$$\mathbb{Z} = \left\{ \begin{matrix} \bullet & * & * & * & \\ * & \bullet & * & * & \bullet \\ * & * & * & * & \bullet \\ * & * & \bullet & * & * \\ * & \bullet & \bullet & * & * \end{matrix} \right\}$$

$\mathbf{H}(\mathbb{Z}, -)$: 1 3 6 10 15 19 22 24 $\to$
$\mathbf{H}(\mathbb{X}, -)$: 1 3 6 7 7 7 7 7 $\to$
$\mathbf{H}(\mathbb{Y}, -)$: 1 3 6 10 15 17 17 17 $\to$
$\mathbf{H}(A, -)$: 1 3 6 7 7 5 2 0 $\to$.

203].
$$\mathbb{Z} = \left\{ \begin{matrix} * & * & \bullet & * & & \\ \bullet & * & * & * & & \\ * & * & * & * & \bullet & * \\ * & * & \bullet & * & \bullet & * \\ \bullet & * & \bullet & * & * & * \end{matrix} \right\}$$

$\mathbf{H}(\mathbb{Z}, -)$: 1 3 6 10 15 20 24 26 $\to$
$\mathbf{H}(\mathbb{X}, -)$: 1 3 6 7 7 7 7 7 $\to$
$\mathbf{H}(\mathbb{Y}, -)$: 1 3 6 10 15 19 19 19 $\to$
$\mathbf{H}(A, -)$: 1 3 6 7 7 6 2 0 $\to$.

207].
$$\mathbb{Z} = \left\{ \begin{matrix} * & * & * & \bullet & & \\ * & * & \bullet & \bullet & & \\ * & \bullet & \bullet & \bullet & \bullet & \bullet \\ * & * & * & * & * & \\ * & * & * & * & * & * \end{matrix} \right\}$$

$\mathbf{H}(\mathbb{Z}, -)$: 1 3 6 10 15 20 24 26 $\to$
$\mathbf{H}(\mathbb{X}, -)$: 1 3 6 7 8 8 8 8 $\to$
$\mathbf{H}(\mathbb{Y}, -)$: 1 3 6 10 15 17 18 18 $\to$
$\mathbf{H}(A, -)$: 1 3 6 7 8 5 2 0 $\to$.

208]. $\mathbb{Z} = \left\{ \begin{matrix} * & * & * & * & & \\ \bullet & * & * & * & & \\ * & * & * & * & * & * \\ * & * & * & \bullet & * & \bullet \\ \bullet & \bullet & \bullet & \bullet & \bullet & * \end{matrix} \right\}$

$\mathbf{H}(\mathbb{Z},-) \ : \ 1 \quad 3 \quad 6 \quad 10 \quad 15 \quad 20 \quad 24 \quad 26 \quad \to$
$\mathbf{H}(\mathbb{X},-) \ : \ 1 \quad 3 \quad 6 \quad 7 \quad 8 \quad 8 \quad 8 \quad 8 \quad \to$
$\mathbf{H}(\mathbb{Y},-) \ : \ 1 \quad 3 \quad 6 \quad 10 \quad 15 \quad 18 \quad 18 \quad 18 \quad \to$
$\mathbf{H}(A,-) \ : \ 1 \quad 3 \quad 6 \quad 7 \quad 8 \quad 6 \quad 2 \quad 0 \quad \to.$

227]. $\mathbb{Z} = \left\{ \begin{matrix} * & * & & & \\ * & * & & & \\ * & * & * & \bullet & \\ * & * & * & * & \\ * & * & \bullet & \bullet & \\ * & * & \bullet & \bullet & \\ * & \bullet & \bullet & \bullet & \end{matrix} \right\}$

$\mathbf{H}(\mathbb{Z},-) \ : \ 1 \quad 3 \quad 6 \quad 10 \quad 14 \quad 18 \quad 22 \quad 24 \quad \to$
$\mathbf{H}(\mathbb{X},-) \ : \ 1 \quad 3 \quad 6 \quad 8 \quad 8 \quad 8 \quad 8 \quad 8 \quad \to$
$\mathbf{H}(\mathbb{Y},-) \ : \ 1 \quad 3 \quad 6 \quad 10 \quad 12 \quad 14 \quad 16 \quad 16 \quad \to$
$\mathbf{H}(A,-) \ : \ 1 \quad 3 \quad 6 \quad 8 \quad 6 \quad 4 \quad 2 \quad 0 \quad \to.$

232]. $\mathbb{Z} = \left\{ \begin{matrix} \bullet & * & & & \\ * & * & & & \\ * & * & & & \\ * & * & \bullet & * & * \\ * & * & * & * & \bullet \\ * & * & * & \bullet & \bullet \\ * & * & \bullet & \bullet & \bullet \end{matrix} \right\}$

$\mathbf{H}(\mathbb{Z},-) \ : \ 1 \quad 3 \quad 6 \quad 10 \quad 15 \quad 20 \quad 24 \quad 26 \quad \to$
$\mathbf{H}(\mathbb{X},-) \ : \ 1 \quad 3 \quad 6 \quad 8 \quad 8 \quad 8 \quad 8 \quad 8 \quad \to$
$\mathbf{H}(\mathbb{Y},-) \ : \ 1 \quad 3 \quad 6 \quad 10 \quad 14 \quad 16 \quad 18 \quad 18 \quad \to$
$\mathbf{H}(A,-) \ : \ 1 \quad 3 \quad 6 \quad 8 \quad 7 \quad 4 \quad 2 \quad 0 \quad \to.$

233]. $\mathbb{Z} = \left\{ \begin{matrix} \bullet & \bullet & \bullet & \bullet & & \\ \bullet & * & * & * & & \\ * & \bullet & * & * & * & \bullet \\ * & * & * & * & * & * \\ * & * & * & * & * & * \end{matrix} \right\}$

$\mathbf{H}(\mathbb{Z},-) \ : \ 1 \quad 3 \quad 6 \quad 10 \quad 15 \quad 20 \quad 24 \quad 26 \quad \to$
$\mathbf{H}(\mathbb{X},-) \ : \ 1 \quad 3 \quad 6 \quad 8 \quad 8 \quad 8 \quad 8 \quad 8 \quad \to$
$\mathbf{H}(\mathbb{Y},-) \ : \ 1 \quad 3 \quad 6 \quad 10 \quad 14 \quad 17 \quad 18 \quad 18 \quad \to$
$\mathbf{H}(A,-) \ : \ 1 \quad 3 \quad 6 \quad 8 \quad 7 \quad 5 \quad 2 \quad 0 \quad \to.$

234]. $\mathbb{Z} = \left\{ \begin{matrix} \bullet & * & * & * & & \\ \bullet & \bullet & \bullet & \bullet & & \\ * & * & * & \bullet & * & * \\ * & * & * & * & * & \bullet \\ * & * & * & \bullet & * & * \end{matrix} \right\}$

$\mathbf{H}(\mathbb{Z},-) \ : \ 1 \quad 3 \quad 6 \quad 10 \quad 15 \quad 20 \quad 24 \quad 26 \quad \to$
$\mathbf{H}(\mathbb{X},-) \ : \ 1 \quad 3 \quad 6 \quad 8 \quad 8 \quad 8 \quad 8 \quad 8 \quad \to$
$\mathbf{H}(\mathbb{Y},-) \ : \ 1 \quad 3 \quad 6 \quad 10 \quad 14 \quad 18 \quad 18 \quad 18 \quad \to$
$\mathbf{H}(A,-) \ : \ 1 \quad 3 \quad 6 \quad 8 \quad 7 \quad 6 \quad 2 \quad 0 \quad \to.$

238]. $\mathbb{Z} = \left\{ \begin{matrix} \bullet & \bullet & \bullet & * & \\ \bullet & \bullet & \bullet & * & * \\ \bullet & \bullet & * & * & * \\ * & * & * & * & * \\ * & * & * & * & * \end{matrix} \right\}$

$\mathbf{H}(\mathbb{Z},-) \ : \ 1 \quad 3 \quad 6 \quad 10 \quad 15 \quad 19 \quad 22 \quad 24 \quad \to$
$\mathbf{H}(\mathbb{X},-) \ : \ 1 \quad 3 \quad 6 \quad 8 \quad 8 \quad 8 \quad 8 \quad 8 \quad \to$
$\mathbf{H}(\mathbb{Y},-) \ : \ 1 \quad 3 \quad 6 \quad 10 \quad 15 \quad 16 \quad 16 \quad 16 \quad \to$
$\mathbf{H}(A,-) \ : \ 1 \quad 3 \quad 6 \quad 8 \quad 8 \quad 5 \quad 2 \quad 0 \quad \to.$

239]. $\mathbb{Z} = \left\{ \begin{matrix} * & * & * & * & & \\ * & * & * & \bullet & & \\ * & * & \bullet & \bullet & \bullet & \bullet \\ * & * & * & \bullet & \bullet & \bullet \\ * & * & * & * & * & * \end{matrix} \right\}$

$\mathbf{H}(\mathbb{Z},-) \ : \ 1 \quad 3 \quad 6 \quad 10 \quad 15 \quad 20 \quad 24 \quad 26 \quad \to$
$\mathbf{H}(\mathbb{X},-) \ : \ 1 \quad 3 \quad 6 \quad 8 \quad 8 \quad 8 \quad 8 \quad 8 \quad \to$
$\mathbf{H}(\mathbb{Y},-) \ : \ 1 \quad 3 \quad 6 \quad 10 \quad 15 \quad 18 \quad 18 \quad 18 \quad \to$
$\mathbf{H}(A,-) \ : \ 1 \quad 3 \quad 6 \quad 8 \quad 8 \quad 6 \quad 2 \quad 0 \quad \to.$

B. SOCLE DEGREE 6 AND TYPE 2

243]. $\mathbb{Z} = \left\{ \begin{matrix} * & & & \\ * & & & \\ * & & & \\ \bullet & \bullet & \bullet & \bullet \\ \bullet & \bullet & \bullet & \bullet \\ * & * & * & \bullet \\ * & * & * & * \\ * & * & * & * \end{matrix} \right\}$

$\mathbf{H}(\mathbb{Z},-)$	:	1	3	6	10	14	18	21	23	$\rightarrow$
$\mathbf{H}(\mathbb{X},-)$	:	1	3	6	8	9	9	9	9	$\rightarrow$
$\mathbf{H}(\mathbb{Y},-)$	:	1	3	6	10	14	14	14	14	$\rightarrow$
$\mathbf{H}(A,-)$	:	1	3	6	8	9	5	2	0	$\rightarrow$.

244]. $\mathbb{Z} = \left\{ \begin{matrix} * & * & * & * & & \\ * & * & * & * & & \\ \bullet & \bullet & \bullet & * & * & * \\ \bullet & \bullet & \bullet & * & * & * \\ \bullet & \bullet & \bullet & * & * & * \end{matrix} \right\}$

$\mathbf{H}(\mathbb{Z},-)$	:	1	3	6	10	15	20	24	26	$\rightarrow$
$\mathbf{H}(\mathbb{X},-)$	:	1	3	6	8	9	9	9	9	$\rightarrow$
$\mathbf{H}(\mathbb{Y},-)$	:	1	3	6	10	15	17	17	17	$\rightarrow$
$\mathbf{H}(A,-)$	:	1	3	6	8	9	6	2	0	$\rightarrow$.

249]. $\mathbb{Z} = \left\{ \begin{matrix} * & * & \bullet & \bullet & & \\ * & * & \bullet & \bullet & & \\ * & * & \bullet & \bullet & \bullet & * \\ * & * & \bullet & \bullet & * & * \\ * & * & * & \bullet & * & * \end{matrix} \right\}$

$\mathbf{H}(\mathbb{Z},-)$	:	1	3	6	10	15	20	24	26	$\rightarrow$
$\mathbf{H}(\mathbb{X},-)$	:	1	3	6	8	10	10	10	10	$\rightarrow$
$\mathbf{H}(\mathbb{Y},-)$	:	1	3	6	10	15	16	16	16	$\rightarrow$
$\mathbf{H}(A,-)$	:	1	3	6	8	10	6	2	0	$\rightarrow$.

268]. $\mathbb{Z} = \left\{ \begin{matrix} * & & & \\ * & & & \\ * & & & \\ * & * & * & * \\ * & * & * & \bullet \\ * & * & \bullet & \bullet \\ * & * & \bullet & \bullet \\ \bullet & \bullet & \bullet & \bullet \end{matrix} \right\}$

$\mathbf{H}(\mathbb{Z},-)$	:	1	3	6	10	14	18	21	23	$\rightarrow$
$\mathbf{H}(\mathbb{X},-)$	:	1	3	6	9	9	9	9	9	$\rightarrow$
$\mathbf{H}(\mathbb{Y},-)$	:	1	3	6	10	12	13	14	14	$\rightarrow$
$\mathbf{H}(A,-)$	:	1	3	6	9	7	4	2	0	$\rightarrow$.

269]. $\mathbb{Z} = \left\{ \begin{matrix} \bullet & \bullet & \bullet & \bullet & & \\ * & \bullet & \bullet & \bullet & & \\ * & * & \bullet & \bullet & * & * \\ * & * & * & * & * & * \\ * & * & * & * & * & * \end{matrix} \right\}$

$\mathbf{H}(\mathbb{Z},-)$	:	1	3	6	10	15	20	24	26	$\rightarrow$
$\mathbf{H}(\mathbb{X},-)$	:	1	3	6	9	9	9	9	9	$\rightarrow$
$\mathbf{H}(\mathbb{Y},-)$	:	1	3	6	10	13	16	17	17	$\rightarrow$
$\mathbf{H}(A,-)$	:	1	3	6	9	7	5	2	0	$\rightarrow$.

274]. $\mathbb{Z} = \left\{ \begin{matrix} * & \bullet & \bullet & \bullet & & \\ * & * & \bullet & \bullet & & \\ * & * & \bullet & \bullet & \bullet & \bullet \\ * & * & * & * & * & * \\ * & * & * & * & * & * \end{matrix} \right\}$

$\mathbf{H}(\mathbb{Z},-)$	:	1	3	6	10	15	20	24	26	$\rightarrow$
$\mathbf{H}(\mathbb{X},-)$	:	1	3	6	9	9	9	9	9	$\rightarrow$
$\mathbf{H}(\mathbb{Y},-)$	:	1	3	6	10	14	16	17	17	$\rightarrow$
$\mathbf{H}(A,-)$	:	1	3	6	9	8	5	2	0	$\rightarrow$.

275]. $\mathbb{Z} = \left\{ \begin{matrix} \bullet & \bullet & \bullet & \bullet & & \\ * & * & * & * & & \\ * & * & * & * & \bullet & * \\ * & * & \bullet & * & * & * \\ * & \bullet & \bullet & \bullet & * & * \end{matrix} \right\}$

$\mathbf{H}(\mathbb{Z},-)$	:	1	3	6	10	15	20	24	26	$\rightarrow$
$\mathbf{H}(\mathbb{X},-)$	:	1	3	6	9	9	9	9	9	$\rightarrow$
$\mathbf{H}(\mathbb{Y},-)$	:	1	3	6	10	14	17	17	17	$\rightarrow$
$\mathbf{H}(A,-)$	:	1	3	6	9	8	6	2	0	$\rightarrow$.

279].
$$\mathbb{Z} = \left\{\begin{matrix} \bullet & \bullet & \bullet & * & \\ \bullet & \bullet & \bullet & * & * \\ * & \bullet & \bullet & * & * \\ * & * & \bullet & * & * \\ * & * & * & * & * \end{matrix}\right\}$$

$\mathbf{H}(\mathbb{Z},-)$: 1 3 6 10 15 19 22 24 $\to$
$\mathbf{H}(\mathbb{X},-)$: 1 3 6 9 9 9 9 9 $\to$
$\mathbf{H}(\mathbb{Y},-)$: 1 3 6 10 15 15 15 15 $\to$
$\mathbf{H}(A,-)$: 1 3 6 9 9 5 2 0 $\to$.

280].
$$\mathbb{Z} = \left\{\begin{matrix} * & * & \bullet & * & & \\ * & * & * & * & & \\ * & * & * & \bullet & \bullet & * \\ * & \bullet & \bullet & \bullet & \bullet & * \\ \bullet & * & * & \bullet & * & \bullet \end{matrix}\right\}$$

$\mathbf{H}(\mathbb{Z},-)$: 1 3 6 10 15 20 24 26 $\to$
$\mathbf{H}(\mathbb{X},-)$: 1 3 6 9 9 9 9 9 $\to$
$\mathbf{H}(\mathbb{Y},-)$: 1 3 6 10 15 17 17 17 $\to$
$\mathbf{H}(A,-)$: 1 3 6 9 9 6 2 0 $\to$.

285].
$$\mathbb{Z} = \left\{\begin{matrix} \bullet & \bullet & \bullet & * & & \\ \bullet & \bullet & * & * & & \\ \bullet & \bullet & * & * & * & \bullet \\ \bullet & \bullet & * & * & * & * \\ * & * & * & * & * & * \end{matrix}\right\}$$

$\mathbf{H}(\mathbb{Z},-)$: 1 3 6 10 15 20 24 26 $\to$
$\mathbf{H}(\mathbb{X},-)$: 1 3 6 9 10 10 10 10 $\to$
$\mathbf{H}(\mathbb{Y},-)$: 1 3 6 10 15 16 16 16 $\to$
$\mathbf{H}(A,-)$: 1 3 6 9 10 6 2 0 $\to$.

290].
$$\mathbb{Z} = \left\{\begin{matrix} \bullet & \bullet & * & * & & \\ \bullet & \bullet & \bullet & * & & \\ \bullet & \bullet & * & * & * & \\ \bullet & \bullet & * & * & * & \\ \bullet & * & * & * & * & * \end{matrix}\right\}$$

$\mathbf{H}(\mathbb{Z},-)$: 1 3 6 10 15 20 24 26 $\to$
$\mathbf{H}(\mathbb{X},-)$: 1 3 6 9 11 11 11 11 $\to$
$\mathbf{H}(\mathbb{Y},-)$: 1 3 6 10 15 15 15 15 $\to$
$\mathbf{H}(A,-)$: 1 3 6 9 11 6 2 0 $\to$.

295].
$$\mathbb{Z} = \left\{\begin{matrix} * & * & * & & & \\ * & * & * & & & \\ * & * & * & & & \\ \bullet & \bullet & * & \bullet & \bullet & * \\ \bullet & \bullet & \bullet & * & * & \bullet \\ \bullet & \bullet & * & * & \bullet & \bullet \end{matrix}\right\}$$

$\mathbf{H}(\mathbb{Z},-)$: 1 3 6 10 15 21 25 27 $\to$
$\mathbf{H}(\mathbb{X},-)$: 1 3 6 9 12 12 12 12 $\to$
$\mathbf{H}(\mathbb{Y},-)$: 1 3 6 10 15 15 15 15 $\to$
$\mathbf{H}(A,-)$: 1 3 6 9 12 6 2 0 $\to$.

314].
$$\mathbb{Z} = \left\{\begin{matrix} * & & & \\ * & & & \\ * & & & \\ \bullet & \bullet & \bullet & \bullet \\ * & \bullet & \bullet & \bullet \\ * & * & \bullet & \bullet \\ * & * & * & \bullet \\ * & * & * & * \end{matrix}\right\}$$

$\mathbf{H}(\mathbb{Z},-)$: 1 3 6 10 14 18 21 23 $\to$
$\mathbf{H}(\mathbb{X},-)$: 1 3 6 10 10 10 10 10 $\to$
$\mathbf{H}(\mathbb{Y},-)$: 1 3 6 10 11 12 13 13 $\to$
$\mathbf{H}(A,-)$: 1 3 6 10 7 4 2 0 $\to$.

320].
$$\mathbb{Z} = \left\{\begin{matrix} * & & & \\ \bullet & & & \\ * & & & \\ * & * & \bullet & * \\ * & * & * & \bullet \\ \bullet & * & \bullet & \bullet \\ * & * & \bullet & \bullet \\ * & \bullet & \bullet & * \end{matrix}\right\}$$

$\mathbf{H}(\mathbb{Z},-)$: 1 3 6 10 14 18 21 23 $\to$
$\mathbf{H}(\mathbb{X},-)$: 1 3 6 10 10 10 10 10 $\to$
$\mathbf{H}(\mathbb{Y},-)$: 1 3 6 10 12 13 13 13 $\to$
$\mathbf{H}(A,-)$: 1 3 6 10 8 5 2 0 $\to$.

321]. $\mathbb{Z} = \left\{\begin{matrix} \bullet & * & \bullet & \bullet \\ \bullet & \bullet & \bullet & \bullet \\ * & \bullet & * & * & * & * \\ * & \bullet & \bullet & * & * \\ * & * & * & * & * \end{matrix}\right\}$
$\mathbf{H}(\mathbb{Z},-)\ :\ 1\ \ 3\ \ 6\ \ 10\ \ 15\ \ 20\ \ 24\ \ 26\ \ \to$
$\mathbf{H}(\mathbb{X},-)\ :\ 1\ \ 3\ \ 6\ \ 10\ \ 10\ \ 10\ \ 10\ \ 10\ \ \to$
$\mathbf{H}(\mathbb{Y},-)\ :\ 1\ \ 3\ \ 6\ \ 10\ \ 13\ \ 16\ \ 16\ \ 16\ \ \to$
$\mathbf{H}(A,-)\ \ :\ 1\ \ 3\ \ 6\ \ 10\ \ \ 8\ \ \ 6\ \ \ 2\ \ \ 0\ \ \to.$

325]. $\mathbb{Z} = \left\{\begin{matrix} \bullet & \bullet & \bullet & \bullet \\ \bullet & \bullet & \bullet & * & * \\ \bullet & \bullet & * & * & * \\ \bullet & * & * & * & * \\ * & * & * & * & * \end{matrix}\right\}$
$\mathbf{H}(\mathbb{Z},-)\ :\ 1\ \ 3\ \ 6\ \ 10\ \ 15\ \ 19\ \ 22\ \ 24\ \ \to$
$\mathbf{H}(\mathbb{X},-)\ :\ 1\ \ 3\ \ 6\ \ 10\ \ 10\ \ 10\ \ 10\ \ 10\ \ \to$
$\mathbf{H}(\mathbb{Y},-)\ :\ 1\ \ 3\ \ 6\ \ 10\ \ 14\ \ 14\ \ 14\ \ 14\ \ \to$
$\mathbf{H}(A,-)\ \ :\ 1\ \ 3\ \ 6\ \ 10\ \ \ 9\ \ \ 5\ \ \ 2\ \ \ 0\ \ \to.$

326]. $\mathbb{Z} = \left\{\begin{matrix} \bullet & \bullet & \bullet & \bullet \\ \bullet & * & * & * \\ \bullet & \bullet & \bullet & * & * & * \\ * & * & * & \bullet & * & * \\ * & * & \bullet & * & * & * \end{matrix}\right\}$
$\mathbf{H}(\mathbb{Z},-)\ :\ 1\ \ 3\ \ 6\ \ 10\ \ 15\ \ 20\ \ 24\ \ 26\ \ \to$
$\mathbf{H}(\mathbb{X},-)\ :\ 1\ \ 3\ \ 6\ \ 10\ \ 10\ \ 10\ \ 10\ \ 10\ \ \to$
$\mathbf{H}(\mathbb{Y},-)\ :\ 1\ \ 3\ \ 6\ \ 10\ \ 14\ \ 16\ \ 16\ \ 16\ \ \to$
$\mathbf{H}(A,-)\ \ :\ 1\ \ 3\ \ 6\ \ 10\ \ \ 9\ \ \ 6\ \ \ 2\ \ \ 0\ \ \to.$

331]. $\mathbb{Z} = \left\{\begin{matrix} \bullet & \bullet & \bullet & * \\ \bullet & \bullet & * & * \\ \bullet & \bullet & * & * & * & \bullet \\ \bullet & * & * & * & * & \bullet \\ * & * & * & * & * \end{matrix}\right\}$
$\mathbf{H}(\mathbb{Z},-)\ :\ 1\ \ 3\ \ 6\ \ 10\ \ 15\ \ 20\ \ 24\ \ 26\ \ \to$
$\mathbf{H}(\mathbb{X},-)\ :\ 1\ \ 3\ \ 6\ \ 10\ \ 10\ \ 10\ \ 10\ \ 10\ \ \to$
$\mathbf{H}(\mathbb{Y},-)\ :\ 1\ \ 3\ \ 6\ \ 10\ \ 15\ \ 16\ \ 16\ \ 16\ \ \to$
$\mathbf{H}(A,-)\ \ :\ 1\ \ 3\ \ 6\ \ 10\ \ 10\ \ \ 6\ \ \ 2\ \ \ 0\ \ \to.$

336]. $\mathbb{Z} = \left\{\begin{matrix} * & * & * & * \\ * & * & * & \bullet \\ * & * & * & \bullet & \bullet & * \\ * & * & \bullet & \bullet & \bullet & * \\ * & \bullet & \bullet & \bullet & \bullet & \bullet \end{matrix}\right\}$
$\mathbf{H}(\mathbb{Z},-)\ :\ 1\ \ 3\ \ 6\ \ 10\ \ 15\ \ 20\ \ 24\ \ 26\ \ \to$
$\mathbf{H}(\mathbb{X},-)\ :\ 1\ \ 3\ \ 6\ \ 10\ \ 11\ \ 11\ \ 11\ \ 11\ \ \to$
$\mathbf{H}(\mathbb{Y},-)\ :\ 1\ \ 3\ \ 6\ \ 10\ \ 15\ \ 15\ \ 15\ \ 15\ \ \to$
$\mathbf{H}(A,-)\ \ :\ 1\ \ 3\ \ 6\ \ 10\ \ 11\ \ \ 6\ \ \ 2\ \ \ 0\ \ \to.$

341]. $\mathbb{Z} = \left\{\begin{matrix} * & * & * \\ * & * & * \\ * & * & \bullet \\ * & * & \bullet & \bullet & * & * \\ * & \bullet & \bullet & \bullet & \bullet & * \\ \bullet & \bullet & \bullet & \bullet & \bullet & * \end{matrix}\right\}$
$\mathbf{H}(\mathbb{Z},-)\ :\ 1\ \ 3\ \ 6\ \ 10\ \ 15\ \ 21\ \ 25\ \ 27\ \ \to$
$\mathbf{H}(\mathbb{X},-)\ :\ 1\ \ 3\ \ 6\ \ 10\ \ 12\ \ 12\ \ 12\ \ 12\ \ \to$
$\mathbf{H}(\mathbb{Y},-)\ :\ 1\ \ 3\ \ 6\ \ 10\ \ 15\ \ 15\ \ 15\ \ 15\ \ \to$
$\mathbf{H}(A,-)\ \ :\ 1\ \ 3\ \ 6\ \ 10\ \ 12\ \ \ 6\ \ \ 2\ \ \ 0\ \ \to.$

Points

We now show that all the h-vectors in **Table 6.2** are also the h-vectors of a level set of points in $\mathbb{P}^3$. This gives a positive answer (in this case) to our **Question 6.1** as to whether every h-vector of an Artinian level algebra in codimension n is also the h-vector of a level set of points in $\mathbb{P}^n$.

In the body of the paper we have already given some examples for each of the four methods we discussed.

Method 1. The higher dimensional "linked-sum" method.

From **Example 6.8** we get that 36] is the h-vector of a level set of points in $\mathbb{P}^3$. We can continue with Method 1, and the diagrams we used above, to also show that: 5], 25], 26], 40], 41], 46], 71], 72], 87], 92], 93], 98], 113], 114], 119], 125], 175],

201], 207], 227], 238], 239], 243], 249], 268], 269], 279], 314], 325] are the h-vectors of level sets of points in $\mathbb{P}^3$.

Method 2. Liaison method.

From **Example 6.10** we get 341]. From **Remark 6.12**, 2) we get 234], using this method.

All the remaining examples we get will pretty much follow the same track as outlined in **Example 6.10** above.

94]. Let $R = k[x_0, x_1, x_2, x_3]$ and consider 7 general points, $\mathbb{Z}_1$, in $\mathbb{P}^3$. The h-vector or $R/I_{\mathbb{Z}_1}$ is $(1,3,3)$ and $I_{\mathbb{Z}_1}$ has minimal free resolution

$$0 \to R(-5)^3 \to R(-4)^6 \to \begin{array}{c} R(-2)^3 \\ \oplus \\ R(-3) \end{array} \to I_{\mathbb{Z}_1} \to 0.$$

Choose a complete intersection of type $(2,3,6)$ inside $I_{\mathbb{Z}_1}$ and call the ideal of the complete intersection J_1. This links the points $\mathbb{Z}_1$ to 29 points ($\mathbb{Z}_2$) in $\mathbb{P}^3$ for which $R/I_{\mathbb{Z}_2}$ has h-vector given by 94]. It remains to show that these 29 points are level.

As in **Example 6.10**, we have

$$0 \to J_1 \to I_{\mathbb{Z}_1} \to K_{\mathbb{Z}_2}(-7) \to 0$$

and, following the procedure of **Example 6.10**, we get that a minimal free resolution of $I_{\mathbb{Z}_2}$ is:

$$0 \to R(-9)^2 \to \begin{array}{c} R(-5) \\ \oplus \\ R(-7)^6 \end{array} \to \begin{array}{c} R(-2) \\ \oplus \\ R(-3) \\ \oplus \\ R(-6)^4 \end{array} \to I_{\mathbb{Z}_2} \to 0.$$

This resolution shows that $\mathbb{Z}_2$ is a level set of points.

99]. This is more complicated but uses the same ideas as above. We start with 1 point and perform general links of types (respectively): $(1,3,3)$, $(2,3,5)$, $(2,4,5)$, $(2,4,6)$ and $(2,5,6)$. We end up with a set of points $\mathbb{Z}$ with the appropriate h-vector and with the (not necessarily minimal) free resolution:

$$0 \to R(-9)^2 \to \begin{array}{c} R(-4) \\ \oplus \\ R(-6)^2 \\ \oplus \\ R(-7)^5 \end{array} \to \begin{array}{c} R(-2) \\ \oplus \\ R(-3) \\ \oplus \\ R(-5)^2 \\ \oplus \\ R(-6)^3 \end{array} \to I_{\mathbb{Z}} \to 0$$

which shows that $\mathbb{Z}$ is a level set of points.

For the remaining examples we will just identify the links that can be made and leave the details to the interested reader.

124]. Start with 2 points and perform general links of types (respectively): $(1,3,3)$, $(2,3,3)$, $(2,3,5)$ and $(2,5,5)$.

203]. Start with 1 point and perform general links of types (respectively): $(1,2,4)$, $(2,2,5)$ and $(3,3,5)$.

B. SOCLE DEGREE 6 AND TYPE 2   91

208]. Start with 1 point and perform general links of types (respectively): (1,3,3), (2,3,5), (2,4,5), (3,4,5) and (3,5,5).

243]. Start with 2 points and perform general links of types (respectively): (1,3,3), (2,3,3), (2,3,4), (2,4,4), (3,4,5) and (3,5,5).

295]. Start with 12 general points and perform general links of types (respectively): (3,4,4) and (3,5,5).

320]. Start with 2 points and perform general links of types (respectively): (1,2,4), (2,2,4), (3,3,4), (3,3,5), (4,4,5) and (4,4,6).

321]. Start with 3 collinear points and perform general links of types (respectively): (1,2,4), (2,2,4), (3,3,4), (3,3,5), (4,4,5) and (4,4,6).

326]. Start with 9 general points on the twisted cubic and perform general links of types (respectively): (3,3,4) and (4,4,4).

331]. Start with a complete intersection set of points of type (2,2,2) and perform general links of types (respectively): (2,3,3), (3,3,4) and (4,4,4).

336]. Start with 7 general points and perform general links of types (respectively): (2,3,3), (3,3,4) and (4,4,4).

Method 3. Unions

These are all obtained by using **Lemma 6.13** and **Corollary 6.14**. We will just indicate the method and leave the verifications to the interested reader.

120]. Use **Lemma 6.13** starting with a set of 11 general points on a twisted cubic in $\mathbb{P}^3$ (which is an arithmetically Gorenstein set of points since it has symmetric h-vector and the points satisfy the Cayley-Bacharach property - see [**15**]) and then apply the Lemma with $d = 2$, $3 = 5$, $f = 2$.

130]. Start with an arithmetically Gorenstein set of points in $\mathbb{P}^3$ with h-vector $(1, 3, 5, 3, 1)$ and apply **Lemma 6.13** with $d = 2$, $e = 5$, $f = 2$.

176]. Start with 14 general points on a twisted cubic in $\mathbb{P}^3$ (an arithmetically Gorenstein set of points for the same reason as in 120]) and apply **Lemma 6.13** with $d = 3$, $e = 5$, $f = 1$.

196]. Use **Corollary 6.14** with $a = b = 1$, $c = 5$, $d = 3$, $e = 4$, $f = 2$.

197]. Start with 14 general points on an twisted cubic in $\mathbb{P}^3$ and apply **Lemma 6.13** with $d = 3 = 4$, $f = 1$.

202]. Use **Corollary 6.14** with $a = b = 2$, $c = 4$, $d = 3$, $e = 5$, $f = 1$.

232]. Use **Corollary 6.14** with $a = b = 1$, $c = 4$, $d = e = f = 3$.

233]. Use **Corollary 6.14** with $a = b = 2$, $c = 4$, $d = e = 4$, $f = 1$.

244]. Use **Corollary 6.14** with $a = b = c = 2$, $d = e = f = 3$.

274]. Use **Corollary 6.14** with $a = 2$, $b = c = 3$, $c = e = 4$, $f = 1$.

275]. Start with 11 general points on the twisted cubic in $\mathbb{P}^3$ and then apply **Lemma 6.13** with $d = 3$, $e = 4$, $f = 2$.

280]. Use **Corollary 6.14** with $a = b = 2$, $c = d = 3$, $e = 4$, $f = 2$.

285]. Start with an arithmetically Gorenstein set of points in $\mathbb{P}^3$ with h-vector $(1, 3, 5, 3, 1)$ and then apply **Lemma 6.13** with $d = 3$, $e = 4$. $f = 2$.

290]. Start with an arithmetically Gorenstein set of points in $\mathbb{P}^3$ with h-vector $(1,3,6,3,1)$ and then apply **Lemma 6.13** with $d=3$, $e=4$, $f=2$.

Method 4. Points on Curves

The only case left to consider is 177], and this can be obtained by considering 30 general points on an arithmetically Cohen-Macaulay curve in $\mathbb{P}^3$ of degree 6 and arithmetic genus 3.

The Weak Lefschetz Property

Each of the examples of a level algebra constructed above, using the Linked-Sum Method, in fact already gives an Artinian algebra with the WLP. Indeed, each of those examples gives a ring A which is a quotient of $B = R/I_\mathbb{X}$ where A and B satisfy the hypotheses of **Proposition 5.15**.

APPENDIX C

Socle Degree 5

Socle Degree 5 and Type 2

We begin by considering what can be the h-vector of a level sequence of type 2 (the Gorenstein case is well-known). In our paper, [**23**] we listed all the h-vectors which can be the h-vector of a level sequence of type 2 and socle degree 5, and promised to put our calculations in this appendix.

The following table lists all the 79 h-vectors of Artinian algebras of socle degree 5 which have last value 2. We start by eliminating from this list all the h-vectors which cannot be the h-vector of a level algebra of type 2.

Table 5.2A

1]	1, 3, 2, 2, 2, 2	2]	1, 3, 3, 2, 2, 2	3]	1, 3, 3, 3, 2, 2
4]	1, 3, 3, 3, 3, 2	5]	1, 3, 3, 4, 2, 2	6]	1, 3, 3, 4, 3, 2
7]	1, 3, 3, 4, 4, 2	8]	1, 3, 3, 4, 5, 2	9]	1, 3, 4, 2, 2, 2
10]	1, 3, 4, 3, 2, 2	11]	1, 3, 4, 3, 3, 2	12]	1, 3, 4, 4, 2, 2
13]	1, 3, 4, 4, 3, 2	14]	1, 3, 4, 4, 4, 2	15]	1, 3, 4, 4, 5, 2
16]	1, 3, 4, 5, 2, 2	17]	1, 3, 4, 5, 3, 2	18]	1, 3, 4, 5, 4, 2
19]	1, 3, 4, 5, 5, 2	20]	1, 3, 4, 5, 6, 2	21]	1, 3, 5, 2, 2, 2
22]	1, 3, 5, 3, 2, 2	23]	1, 3, 5, 3, 3, 2	24]	1, 3, 5, 4, 2, 2
25]	1, 3, 5, 4, 3, 2	26]	1, 3, 5, 4, 4, 2	27]	1, 3, 5, 4, 5, 2
28]	1, 3, 5, 5, 2, 2	29]	1, 3, 5, 5, 3, 2	30]	1, 3, 5, 5, 4, 2
31]	1, 3, 5, 5, 5, 2	32]	1, 3, 5, 5, 6, 2	33]	1, 3, 5, 6, 2, 2
34]	1, 3, 5, 6, 3, 2	35]	1, 3, 5, 6, 4, 2	36]	1, 3, 5, 6, 5, 2
37]	1, 3, 5, 6, 6, 2	38]	1, 3, 5, 7, 2, 2	39]	1, 3, 5, 7, 3, 2
40]	1, 3, 5, 7, 4, 2	41]	1, 3, 5, 7, 5, 2	42]	1, 3, 5, 7, 6, 2
43]	1, 3, 6, 2, 2, 2	44]	1, 3, 6, 3, 2, 2	45]	1, 3, 6, 3, 3, 2
46]	1, 3, 6, 4, 2, 2	47]	1, 3, 6, 4, 3, 2	48]	1, 3, 6, 4, 4, 2
49]	1, 3, 6, 4, 5, 2	50]	1, 3, 6, 5, 2, 2	51]	1, 3, 6, 5, 3, 2
52]	1, 3, 6, 5, 4, 2	53]	1, 3, 6, 5, 5, 2	54]	1, 3, 6, 5, 6, 2
55]	1, 3, 6, 6, 2, 2	56]	1, 3, 6, 6, 3, 2	57]	1, 3, 6, 6, 4, 2
58]	1, 3, 6, 6, 5, 2	59]	1, 3, 6, 6, 6, 2	60]	1, 3, 6, 7, 2, 2
61]	1, 3, 6, 7, 3, 2	62]	1, 3, 6, 7, 4, 2	63]	1, 3, 6, 7, 5, 2
64]	1, 3, 6, 7, 6, 2	65]	1, 3, 6, 8, 2, 2	66]	1, 3, 6, 8, 3, 2
67]	1, 3, 6, 8, 4, 2	68]	1, 3, 6, 8, 5, 2	69]	1, 3, 6, 8, 6, 2
70]	1, 3, 6, 9, 2, 2	71]	1, 3, 6, 9, 3, 2	72]	1, 3, 6, 9, 4, 2
73]	1, 3, 6, 9, 5, 2	74]	1, 3, 6, 9, 6, 2	75]	1, 3, 6, 10, 2, 2
76]	1, 3, 6, 10, 3, 2	77]	1, 3, 6, 10, 4, 2	78]	1, 3, 6, 10, 5, 2
79]	1, 3, 6, 10, 6, 2				

Non-existence

Eliminated by Remark 2.12, b). $(\mathbf{1,3,...,2,2})$: 1], 2], 3], 5], 9], 10], 12], 16], 21], 22], 24], 28], 33], 38], 43], 44], 46], 50], 55], 60], 65], 70], 75].

$(\mathbf{1,3,\ldots,8,4,2}), (\mathbf{1,3,\ldots,9,4,2}), (\mathbf{1,3,\ldots,10,4,2}), (\mathbf{1,3,\ldots,10,5,2})$: 67], 72], 77], 78].

$(\mathbf{1,3,6,4,3,2})$: 47].

Eliminated by Example 2.13. 17], 29], 51].

Eliminated by Corollary 3.5. 6], 7], 8], 15], 27], 32], 49], 54].
Eliminated by Proposition 3.8.
Part b): for $d = 3$: 11], 23], 45].
Part c): for $d = 3$: 26], 48].

Eliminated by Proposition 3.9. 34], 39], 56], 61], 66], 71], 76].

Eliminated by Example 5.7. 52].

Eliminated by Example 3.11. In this case, we mean by the same argument as in **Example 3.11**. 53].

Eliminated by Corollary 2.11.. 25], 40].

Not level, but not by using any of our theorems. 20] (see also Example 7.3 and [[**14**], Proposition 2.7]).

Existence

We now turn to the existence question. We show that the remaining 23 h-vectors are all the h-vectors of an Artinian level algebra of socle degree 5 and type 2. Those 23 h-vectors are:

Table 5.2

4]	1, 3, 3, 3, 3, 2	13]	1, 3, 4, 4, 3, 2	14]	1, 3, 4, 4, 4, 2
18]	1, 3, 4, 5, 4, 2	19]	1, 3, 4, 5, 5, 2	30]	1, 3, 5, 5, 4, 2
31]	1, 3, 5, 5, 5, 2	35]	1, 3, 5, 6, 4, 2	36]	1, 3, 5, 6, 5, 2
37]	1, 3, 5, 6, 6, 2	41]	1, 3, 5, 7, 5, 2	42]	1, 3, 5, 7, 6, 2
57]	1, 3, 6, 6, 4, 2	58]	1, 3, 6, 6, 5, 2	59]	1, 3, 6, 6, 6, 2
62]	1, 3, 6, 7, 4, 2	63]	1, 3, 6, 7, 5, 2	64]	1, 3, 6, 7, 6, 2
68]	1, 3, 6, 8, 5, 2	69]	1, 3, 6, 8, 6, 2	73]	1, 3, 6, 9, 5, 2
74]	1, 3, 6, 9, 6, 2	79]	1, 3, 6, 10, 6, 2		

Link-Sum Construction. We can use the Linked-Sum Method to construct Artinian level algebras of socle degree and type 2 with each of the above h-vectors. It is clear, from **Proposition 5.15**, that each example also satisfies the WLP. We now do that: in all cases, let $\mathbb{X}$ be the set of all •'s in $\mathbb{Z}$ and $\mathbb{Y}$ be the set of all $*$'s in $\mathbb{Z}$. The ring $A = R/(I_{\mathbb{X}} + I_{\mathbb{Y}})$ is the level algebra with the desired h-vector

4]. $\mathbb{Z} = \left\{\begin{matrix} * & \bullet & \\ \bullet & * & \bullet \\ * & * & * \\ * & * & * \\ * & * & * \end{matrix}\right\}$
$\mathbf{H}(\mathbb{Z}, -)$: 1 3 6 9 12 15 17 $\to$
$\mathbf{H}(\mathbb{X}, -)$: 1 3 3 3 3 3 3 $\to$
$\mathbf{H}(\mathbb{Y}, -)$: 1 3 6 9 12 14 14 $\to$
$\mathbf{H}(A, -)$: 1 3 3 3 3 2 0 $\to$.

13]. $\mathbb{Z} = \left\{\begin{matrix} * & * & \\ * & * & * \\ * & * & * \\ * & * & \bullet \\ * & * & \bullet \\ * & \bullet & \bullet \end{matrix}\right\}$
$\mathbf{H}(\mathbb{Z}, -)$: 1 3 6 9 12 15 17 $\to$
$\mathbf{H}(\mathbb{X}, -)$: 1 3 4 4 4 4 4 $\to$
$\mathbf{H}(\mathbb{Y}, -)$: 1 3 6 9 11 13 13 $\to$
$\mathbf{H}(A, -)$: 1 3 4 4 3 2 0 $\to$.

14]. $\mathbb{Z} = \left\{\begin{matrix} * & \bullet & \\ \bullet & * & \bullet \\ * & \bullet & * \\ * & * & * \\ * & * & * \\ * & * & * \end{matrix}\right\}$
$\mathbf{H}(\mathbb{Z}, -)$: 1 3 6 9 12 15 17 $\to$
$\mathbf{H}(\mathbb{X}, -)$: 1 3 4 4 4 4 4 $\to$
$\mathbf{H}(\mathbb{Y}, -)$: 1 3 6 9 12 13 13 $\to$
$\mathbf{H}(A, -)$: 1 3 4 4 4 2 0 $\to$.

18]. $\mathbb{Z} = \left\{\begin{matrix} \bullet & \bullet & \\ \bullet & * & * \\ \bullet & * & * \\ * & * & * \\ * & * & * \end{matrix}\right\}$
$\mathbf{H}(\mathbb{Z}, -)$: 1 3 6 9 12 15 17 $\to$
$\mathbf{H}(\mathbb{X}, -)$: 1 3 4 5 5 5 5 $\to$
$\mathbf{H}(\mathbb{Y}, -)$: 1 3 6 9 11 12 12 $\to$
$\mathbf{H}(A, -)$: 1 3 4 5 4 2 0 $\to$.

19]. $\mathbb{Z} = \left\{\begin{matrix} * & * & & \\ * & * & & \\ * & * & * & * \\ * & * & * & * \\ * & * & * & \bullet \\ \bullet & \bullet & \bullet & \bullet \end{matrix}\right\}$
$\mathbf{H}(\mathbb{Z}, -)$: 1 3 6 10 14 18 20 $\to$
$\mathbf{H}(\mathbb{X}, -)$: 1 3 4 5 5 5 5 $\to$
$\mathbf{H}(\mathbb{Y}, -)$: 1 3 6 10 14 15 15 $\to$
$\mathbf{H}(A, -)$: 1 3 4 5 5 2 0 $\to$.

30]. $\mathbb{Z} = \left\{\begin{matrix} \bullet & * & \\ * & * & * \\ * & * & * \\ * & * & \bullet \\ * & * & \bullet \\ \bullet & * & \bullet \end{matrix}\right\}$
$\mathbf{H}(\mathbb{Z}, -)$: 1 3 6 9 12 15 17 $\to$
$\mathbf{H}(\mathbb{X}, -)$: 1 3 5 5 5 5 5 $\to$
$\mathbf{H}(\mathbb{Y}, -)$: 1 3 6 9 11 12 12 $\to$
$\mathbf{H}(A, -)$: 1 3 5 5 4 2 0 $\to$.

31]. $\mathbb{Z} = \left\{\begin{matrix} \bullet & \bullet & \\ \bullet & \bullet & * \\ \bullet & * & * \\ * & * & * \\ * & * & * \\ * & * & * \end{matrix}\right\}$
$\mathbf{H}(\mathbb{Z}, -)$: 1 3 6 9 12 15 17 $\to$
$\mathbf{H}(\mathbb{X}, -)$: 1 3 5 5 5 5 5 $\to$
$\mathbf{H}(\mathbb{Y}, -)$: 1 3 6 9 12 12 12 $\to$
$\mathbf{H}(A, -)$: 1 3 5 5 5 2 0 $\to$.

35]. $\mathbb{Z} = \left\{\begin{matrix} * & * & \\ * & * & * \\ \bullet & * & * \\ \bullet & * & * \\ \bullet & * & \bullet \\ \bullet & * & \bullet \end{matrix}\right\}$
$\mathbf{H}(\mathbb{Z},-)\ :\ 1\ \ 3\ \ 6\ \ \ 9\ \ \ 12\ \ \ 15\ \ \ 17\ \to$
$\mathbf{H}(\mathbb{X},-)\ :\ 1\ \ 3\ \ 5\ \ \ 6\ \ \ \ 6\ \ \ \ 6\ \ \ \ 6\ \to$
$\mathbf{H}(\mathbb{Y},-)\ :\ 1\ \ 3\ \ 6\ \ \ 9\ \ \ 10\ \ \ 11\ \ \ 11\ \to$
$\mathbf{H}(A,-)\ :\ 1\ \ 3\ \ 5\ \ \ 6\ \ \ \ 4\ \ \ \ 2\ \ \ \ 0\ \to\,.$

36]. $\mathbb{Z} = \left\{\begin{matrix} * & * & \\ * & * & * \\ \bullet & * & * \\ \bullet & * & * \\ \bullet & \bullet & * \\ \bullet & \bullet & * \end{matrix}\right\}$
$\mathbf{H}(\mathbb{Z},-)\ :\ 1\ \ 3\ \ 6\ \ \ 9\ \ \ 12\ \ \ 15\ \ \ 17\ \to$
$\mathbf{H}(\mathbb{X},-)\ :\ 1\ \ 3\ \ 5\ \ \ 6\ \ \ \ 6\ \ \ \ 6\ \ \ \ 6\ \to$
$\mathbf{H}(\mathbb{Y},-)\ :\ 1\ \ 3\ \ 6\ \ \ 9\ \ \ 11\ \ \ 11\ \ \ 11\ \to$
$\mathbf{H}(A,-)\ :\ 1\ \ 3\ \ 5\ \ \ 6\ \ \ \ 5\ \ \ \ 2\ \ \ \ 0\ \to\,.$

37]. $\mathbb{Z} = \left\{\begin{matrix} * & \bullet & * & & \\ * & * & * & & \\ \bullet & * & * & * & * \\ \bullet & \bullet & * & * & * \\ \bullet & \bullet & * & * & * \end{matrix}\right\}$
$\mathbf{H}(\mathbb{Z},-)\ :\ 1\ \ 3\ \ 6\ \ \ 10\ \ \ 15\ \ \ 19\ \ \ 21\ \to$
$\mathbf{H}(\mathbb{X},-)\ :\ 1\ \ 3\ \ 5\ \ \ 6\ \ \ \ 6\ \ \ \ 6\ \ \ \ 6\ \to$
$\mathbf{H}(\mathbb{Y},-)\ :\ 1\ \ 3\ \ 6\ \ \ 10\ \ \ 15\ \ \ 15\ \ \ 15\ \to$
$\mathbf{H}(A,-)\ :\ 1\ \ 3\ \ 5\ \ \ 6\ \ \ \ 6\ \ \ \ 2\ \ \ \ 0\ \to\,.$

41]. $\mathbb{Z} = \left\{\begin{matrix} * & * & \\ * & * & * \\ \bullet & * & * \\ \bullet & \bullet & * \\ \bullet & \bullet & * \\ \bullet & \bullet & * \end{matrix}\right\}$
$\mathbf{H}(\mathbb{Z},-)\ :\ 1\ \ 3\ \ 6\ \ \ 9\ \ \ 12\ \ \ 15\ \ \ 17\ \to$
$\mathbf{H}(\mathbb{X},-)\ :\ 1\ \ 3\ \ 5\ \ \ 7\ \ \ \ 7\ \ \ \ 7\ \ \ \ 7\ \to$
$\mathbf{H}(\mathbb{Y},-)\ :\ 1\ \ 3\ \ 6\ \ \ 9\ \ \ 10\ \ \ 10\ \ \ 10\ \to$
$\mathbf{H}(A,-)\ :\ 1\ \ 3\ \ 5\ \ \ 7\ \ \ \ 5\ \ \ \ 2\ \ \ \ 0\ \to\,.$

42]. $\mathbb{Z} = \left\{\begin{matrix} * & * & & \\ \bullet & * & & \\ * & \bullet & * & * \\ * & \bullet & \bullet & * \\ * & * & \bullet & \bullet \\ * & * & * & \bullet \end{matrix}\right\}$
$\mathbf{H}(\mathbb{Z},-)\ :\ 1\ \ 3\ \ 6\ \ \ 10\ \ \ 14\ \ \ 18\ \ \ 20\ \to$
$\mathbf{H}(\mathbb{X},-)\ :\ 1\ \ 3\ \ 5\ \ \ 7\ \ \ \ 7\ \ \ \ 7\ \ \ \ 7\ \to$
$\mathbf{H}(\mathbb{Y},-)\ :\ 1\ \ 3\ \ 6\ \ \ 10\ \ \ 13\ \ \ 13\ \ \ 13\ \to$
$\mathbf{H}(A,-)\ :\ 1\ \ 3\ \ 5\ \ \ 7\ \ \ \ 6\ \ \ \ 2\ \ \ \ 0\ \to\,.$

57]. $\mathbb{Z} = \left\{\begin{matrix} * & & \\ * & & \\ * & * & * \\ \bullet & * & \bullet \\ * & \bullet & \bullet \\ * & \bullet & \bullet \\ * & \bullet & * \end{matrix}\right\}$
$\mathbf{H}(\mathbb{Z},-)\ :\ 1\ \ 3\ \ 6\ \ \ 9\ \ \ 12\ \ \ 15\ \ \ 17\ \to$
$\mathbf{H}(\mathbb{X},-)\ :\ 1\ \ 3\ \ 6\ \ \ 6\ \ \ \ 6\ \ \ \ 6\ \ \ \ 6\ \to$
$\mathbf{H}(\mathbb{Y},-)\ :\ 1\ \ 3\ \ 6\ \ \ 9\ \ \ 10\ \ \ 11\ \ \ 11\ \to$
$\mathbf{H}(A,-)\ :\ 1\ \ 3\ \ 6\ \ \ 6\ \ \ \ 4\ \ \ \ 2\ \ \ \ 0\ \to\,.$

58]. $\mathbb{Z} = \left\{\begin{matrix} \bullet & \bullet & \\ \bullet & \bullet & \bullet \\ \bullet & * & * \\ * & * & * \\ * & * & * \\ * & * & * \end{matrix}\right\}$
$\mathbf{H}(\mathbb{Z},-)\ :\ 1\ \ 3\ \ 6\ \ \ 9\ \ \ 12\ \ \ 15\ \ \ 17\ \to$
$\mathbf{H}(\mathbb{X},-)\ :\ 1\ \ 3\ \ 6\ \ \ 6\ \ \ \ 6\ \ \ \ 6\ \ \ \ 6\ \to$
$\mathbf{H}(\mathbb{Y},-)\ :\ 1\ \ 3\ \ 6\ \ \ 9\ \ \ 11\ \ \ 11\ \ \ 11\ \to$
$\mathbf{H}(A,-)\ :\ 1\ \ 3\ \ 6\ \ \ 6\ \ \ \ 5\ \ \ \ 2\ \ \ \ 0\ \to\,.$

C. SOCLE DEGREE 5

59]. $\mathbb{Z} = \left\{\begin{matrix} * & \bullet & \bullet & & \\ * & * & \bullet & & \\ * & * & * & \bullet & \bullet \\ * & * & * & * & * \\ * & * & * & * & \bullet \end{matrix}\right\}$
$\mathbf{H}(\mathbb{Z},-)$: 1 3 6 10 15 19 21 $\to$
$\mathbf{H}(\mathbb{X},-)$: 1 3 6 6 6 6 6 $\to$
$\mathbf{H}(\mathbb{Y},-)$: 1 3 6 10 15 15 15 $\to$
$\mathbf{H}(A,-)$: 1 3 6 6 6 2 0 $\to$.

62]. $\mathbb{Z} = \left\{\begin{matrix} * & & & \\ * & & & \\ * & & & \\ \bullet & * & * & * \\ * & * & \bullet & \bullet \\ * & \bullet & \bullet & * \\ * & \bullet & \bullet & * \end{matrix}\right\}$
$\mathbf{H}(\mathbb{Z},-)$: 1 3 6 10 14 17 19 $\to$
$\mathbf{H}(\mathbb{X},-)$: 1 3 6 7 7 7 7 $\to$
$\mathbf{H}(\mathbb{Y},-)$: 1 3 6 10 11 12 12 $\to$
$\mathbf{H}(A,-)$: 1 3 6 7 4 2 0 $\to$.

63]. $\mathbb{Z} = \left\{\begin{matrix} * & * & \\ * & * & * \\ * & * & * \\ * & * & \bullet \\ \bullet & \bullet & \bullet \\ \bullet & \bullet & \bullet \end{matrix}\right\}$
$\mathbf{H}(\mathbb{Z},-)$: 1 3 6 9 12 15 17 $\to$
$\mathbf{H}(\mathbb{X},-)$: 1 3 6 7 7 7 7 $\to$
$\mathbf{H}(\mathbb{Y},-)$: 1 3 6 9 10 10 10 $\to$
$\mathbf{H}(A,-)$: 1 3 6 7 5 2 0 $\to$.

64]. $\mathbb{Z} = \left\{\begin{matrix} * & * & & \\ * & * & & \\ * & * & * & \\ * & \bullet & * & \bullet \\ * & \bullet & * & \bullet \\ \bullet & \bullet & * & \bullet \end{matrix}\right\}$
$\mathbf{H}(\mathbb{Z},-)$: 1 3 6 10 14 18 20 $\to$
$\mathbf{H}(\mathbb{X},-)$: 1 3 6 7 7 7 7 $\to$
$\mathbf{H}(\mathbb{Y},-)$: 1 3 6 10 13 13 13 $\to$
$\mathbf{H}(A,-)$: 1 3 6 7 6 2 0 $\to$.

68]. $\mathbb{Z} = \left\{\begin{matrix} \bullet & \bullet & \\ \bullet & \bullet & \bullet \\ \bullet & \bullet & * \\ \bullet & * & * \\ * & * & * \\ * & * & * \end{matrix}\right\}$
$\mathbf{H}(\mathbb{Z},-)$: 1 3 6 9 12 15 17 $\to$
$\mathbf{H}(\mathbb{X},-)$: 1 3 6 8 8 8 8 $\to$
$\mathbf{H}(\mathbb{Y},-)$: 1 3 6 9 9 9 9 $\to$
$\mathbf{H}(A,-)$: 1 3 6 8 5 2 0 $\to$.

69]. $\mathbb{Z} = \left\{\begin{matrix} \bullet & \bullet & & \\ \bullet & \bullet & & \\ \bullet & \bullet & \bullet & * \\ \bullet & * & * & * \\ * & * & * & * \\ * & * & * & * \end{matrix}\right\}$
$\mathbf{H}(\mathbb{Z},-)$: 1 3 6 10 14 18 20 $\to$
$\mathbf{H}(\mathbb{X},-)$: 1 3 6 8 8 8 8 $\to$
$\mathbf{H}(\mathbb{Y},-)$: 1 3 6 10 12 12 12 $\to$
$\mathbf{H}(A,-)$: 1 3 6 8 6 2 0 $\to$.

73]. $\mathbb{Z} = \left\{\begin{matrix} \bullet & \bullet & \bullet & \\ \bullet & \bullet & \bullet & * \\ \bullet & \bullet & * & * \\ \bullet & * & * & * \\ * & * & * & * \end{matrix}\right\}$
$\mathbf{H}(\mathbb{Z},-)$: 1 3 6 10 14 17 19 $\to$
$\mathbf{H}(\mathbb{X},-)$: 1 3 6 9 9 9 9 $\to$
$\mathbf{H}(\mathbb{Y},-)$: 1 3 6 10 10 10 10 $\to$
$\mathbf{H}(A,-)$: 1 3 6 9 5 2 0 $\to$.

74]. $\mathbb{Z} = \left\{\begin{matrix} \bullet & \bullet & & \\ \bullet & \bullet & & \\ * & * & \bullet & \bullet \\ * & * & \bullet & \bullet \\ * & * & * & \bullet \\ * & * & * & * \end{matrix}\right\}$
$\mathbf{H}(\mathbb{Z}, -)$: 1 3 6 10 14 18 20 →
$\mathbf{H}(\mathbb{X}, -)$: 1 3 6 9 9 9 9 →
$\mathbf{H}(\mathbb{Y}, -)$: 1 3 6 10 11 11 11 →
$\mathbf{H}(A, -)$: 1 3 6 9 6 2 0 → .

79]. $\mathbb{Z} = \left\{\begin{matrix} \bullet & \bullet & & \\ \bullet & \bullet & & \\ * & \bullet & \bullet & \bullet \\ * & * & * & \bullet \\ * & * & \bullet & \bullet \\ * & * & * & * \end{matrix}\right\}$
$\mathbf{H}(\mathbb{Z}, -)$: 1 3 6 10 14 18 20 →
$\mathbf{H}(\mathbb{X}, -)$: 1 3 6 10 10 10 10 →
$\mathbf{H}(\mathbb{Y}, -)$: 1 3 6 10 10 10 10 →
$\mathbf{H}(A, -)$: 1 3 6 10 6 2 0 → .

Points

We now show that all the h-vectors in **Table 5.2** are also the h-vectors of a level set of points in $\mathbb{P}^3$. This gives a positive answer (in this case) to our **Question 6.1** as to whether every h-vector of an Artinian level algebra in codimension n is also the h-vector of a level set of points in $\mathbb{P}^n$.

Method 1. The higher dimensional "linked-sum" method.

We can use **Example 6.8** and the examples we made above to get that the following are the h-vectors of level sets of points in $\mathbb{P}^3$: 13], 18], 19], 31], 58], 63], 68], 69], 73].

We can also use **Example 6.8** with the following examples. This will give us points for 30], 42], 57], and 62].

30]. $\mathbb{Z} = \left\{\begin{matrix} \bullet & \bullet & \\ \bullet & \bullet & \bullet \\ \bullet & \bullet & * \\ \bullet & \bullet & * \\ * & \bullet & * \\ * & \bullet & * \end{matrix}\right\}$
$\mathbf{H}(\mathbb{Z}, -)$: 1 3 6 9 12 15 17 →
$\mathbf{H}(\mathbb{X}, -)$: 1 3 6 8 10 11 11 →
$\mathbf{H}(\mathbb{Y}, -)$: 1 3 5 6 6 6 6 →
$\mathbf{H}(A, -)$: 1 3 5 5 4 2 0 → .

42]. $\mathbb{Z} = \left\{\begin{matrix} \bullet & \bullet & & \\ \bullet & \bullet & & \\ \bullet & \bullet & \bullet & \bullet \\ \bullet & \bullet & \bullet & \bullet \\ * & * & * & * \\ * & * & * & * \end{matrix}\right\}$
$\mathbf{H}(\mathbb{Z}, -)$: 1 3 6 10 14 18 20 →
$\mathbf{H}(\mathbb{X}, -)$: 1 3 6 10 12 12 12 →
$\mathbf{H}(\mathbb{Y}, -)$: 1 3 5 7 8 8 8 →
$\mathbf{H}(A, -)$: 1 3 5 7 6 2 0 → .

57]. $\mathbb{Z} = \left\{\begin{matrix} \bullet & \bullet & & \\ \bullet & \bullet & & \\ \bullet & \bullet & \bullet & * \\ \bullet & \bullet & * & * \\ \bullet & \bullet & * & * \\ \bullet & * & * & * \end{matrix}\right\}$
$\mathbf{H}(\mathbb{Z}, -)$: 1 3 6 10 14 18 20 →
$\mathbf{H}(\mathbb{X}, -)$: 1 3 6 8 10 12 12 →
$\mathbf{H}(\mathbb{Y}, -)$: 1 3 6 8 8 8 8 →
$\mathbf{H}(A, -)$: 1 3 6 6 4 2 0 → .

62]. $\mathbb{Z} = \left\{\begin{matrix} \bullet & & & \\ \bullet & & & \\ \bullet & & & \\ \bullet & & & \\ \bullet & \bullet & \bullet & * \\ \bullet & \bullet & * & * \\ \bullet & \bullet & * & * \\ * & * & * & * \end{matrix}\right\}$
$\begin{array}{llllllllll} \mathbf{H}(\mathbb{Z},-) & : & 1 & 3 & 6 & 10 & 14 & 17 & 19 & \to \\ \mathbf{H}(\mathbb{X},-) & : & 1 & 3 & 6 & 8 & 9 & 10 & 10 & \to \\ \mathbf{H}(\mathbb{Y},-) & : & 1 & 3 & 6 & 9 & 9 & 9 & 9 & \to \\ \mathbf{H}(A,-) & : & 1 & 3 & 6 & 7 & 4 & 2 & 0 & \to \end{array}.$

Method 2. Liaison method

Using this method we can construct points in $\mathbb{P}^3$ for 35], 36], 37], 41], 64], 74], and 79]. The procedure mostly follows that of **Example 6.10**, although some of the constructions are surprisingly complicated. We indicate, for each example, the procedure we followed.

35]. To obtain this h-vector for points in $\mathbb{P}^3$ one can start with three collinear points and link using a general complete intersection of type $(2,3,4)$.

36]. To obtain this h-vector for points in $\mathbb{P}^3$ one can start with a set of two points and link using a general complete intersection of type $(2,3,4)$.

37]. To obtain this h-vector for points in $\mathbb{P}^3$ one can start with 7 general points and link using a general complete intersection of type $(2,3,5)$.

41]. To obtain this h-vector for points in $\mathbb{P}^3$ one can start with a set, $\mathbb{Z}$, of two points and link using a general arithmetically Gorenstein set of points, G with h-vector $(1,3,5,7,5,3,1)$ containing $\mathbb{Z}$. (Actually one would first choose G and then take two points of G to give $\mathbb{Z}$.) Note that the minimal generators of $I_\mathbb{Z}$ have degrees $1,1,2$, those of I_G have degrees $2,4,4,4,4$ and that the generator of degree 2 splits off with the generator of $I_\mathbb{Z}$ of degree 2 in the mapping cone. (If it did not then the generator of I_G of degree 2 would be a linear combination of the two generators of $I_\mathbb{Z}$ of degree 1, and hence the quadric surface containing G would contain the line spanned by the two points of $\mathbb{Z}$. But without loss of generality we can choose $\mathbb{Z} \subset G$ to be two points whose span does not lie on this surface.

64]. To achieve this for points in $\mathbb{P}^3$, start with one point and apply a sequence of complete intersection links, chosen generally, of the following types (in this order): $(1,2,3), (2,2,4), (2,3,4), (3,4,4), (3,4,5)$.

74]. To achieve this for points in $\mathbb{P}^3$, start with a set of two points and apply a sequence of general complete intersection links of the following types (in this order): $(2,2,2), (2,3,3), (3,3,3), (3,3,4), (3,4,4)$.

79]. To achieve this for points in $\mathbb{P}^3$ is more complicated. We proceed in several steps.

  (a) It is easy to see that this h-vector can be obtained (numerically) as the residual to the h-vector $(1,3,4,4)$ inside a Gorenstein h-vector $(1,3,6,10,10,6,3,1)$. The difficult part is to show that there exist sets of points $\mathbb{Z} \subset G$ with G arithmetically Gorenstein, having these h-vectors.

  (b) A general arithmetically Gorenstein G with the h-vector given in (a) has generators of degree $4,4,4,4,4,5,5,5,5$. We have to show that such a G exists, containing $\mathbb{Z}$. Furthermore, the four quartic generators of $I_\mathbb{Z}$ have to all be split off in the mapping cone. This guarantees that the residual will be level, and liaison gives the right Hilbert function.

(c) Rather than give a long argument for the existence of such a G, we produced it on the computer program *Macaulay* [**2**]. First, choosing $\mathbb{Z}$ is straightforward: one simply starts with 8 general points, finds a complete intersection curve C of type $(2,2)$ containing them, and adds a hyperplane section of C. This gives 12 points that are "general enough".

(d) To produce G containing $\mathbb{Z}$, we used the method of taking sections of Buchsbaum-Rim sheaves [**58**], [**57**] (see in particular [[**58**], Section 5.1] and [[**57**], Section 6]. We start with a sufficiently general matrix of homogeneous polynomials with degree matrix

$$\begin{pmatrix} 1 & 1 & 1 & 1 & 2 & 2 \\ 1 & 1 & 1 & 1 & 2 & 2 \\ 1 & 1 & 1 & 1 & 2 & 2 \end{pmatrix}$$

This defines a map

$$\mathcal{O}_{\mathbb{P}^3}(-1)^4 \oplus \mathcal{O}_{\mathbb{P}^3}(-2)^2 \to \mathcal{O}_{\mathbb{P}^3}^3,$$

whose kernel, $\mathcal{B}$, is the Buchsbaum-Rim sheaf that we will use. Then a general section of $\mathcal{B} \otimes \mathcal{I}_{\mathbb{Z}}(6)$ gives the arithmetically Gorenstein zeroscheme G that contains $\mathbb{Z}$ and has the desired Hilbert function.

Method 4. Points on Curves

With this method we can construct the remaining three cases 4], 14], and 59].

4]. As points in $\mathbb{P}^3$, this can be realized by a set of 15 points on a twisted cubic curve.

14]. As points in $\mathbb{P}^3$, this can be realized as a set of 18 points on a complete intersection curve of type $(2,2)$.

59]. This is realized as points in $\mathbb{P}^3$ by taking 24 general points on a smooth arithmetically Cohen-Macaulay curve of degree 6 and genus 3.

Socle Degree 5 and Type 3

After one eliminates a few trivial cases one is left with a list of 73 tuples which could possibly be the h-vector of an Artinian level algebra of socle degree 5 and type 3. They are in the list below.

Table 5.3A

1]	1, 3, 3, 3, 3, 3	2]	1, 3, 3, 4, 3, 3	3]	1, 3, 3, 4, 4, 3
4]	1, 3, 3, 4, 5, 3	5]	1, 3, 4, 3, 3, 3	6]	1, 3, 4, 4, 3, 3
7]	1, 3, 4, 4, 4, 3	8]	1, 3, 4, 4, 5, 3	9]	1, 3, 4, 5, 3, 3
10]	1, 3, 4, 5, 4, 3	11]	1, 3, 4, 5, 5, 3	12]	1, 3, 4, 5, 6, 3
13]	1, 3, 5, 3, 3, 3	14]	1, 3, 5, 4, 3, 3	15]	1, 3, 5, 4, 4, 3
16]	1, 3, 5, 4, 5, 3	17]	1, 3, 5, 5, 3, 3	18]	1, 3, 5, 5, 4, 3
19]	1, 3, 5, 5, 5, 3	20]	1, 3, 5, 5, 6, 3	21]	1, 3, 5, 6, 3, 3
22]	1, 3, 5, 6, 4, 3	23]	1, 3, 5, 6, 5, 3	24]	1, 3, 5, 6, 6, 3
25]	1, 3, 5, 6, 7, 3	26]	1, 3, 5, 7, 3, 3	27]	1, 3, 5, 7, 4, 3
28]	1, 3, 5, 7, 5, 3	29]	1, 3, 5, 7, 6, 3	30]	1, 3, 5, 7, 7, 3
31]	1, 3, 5, 7, 8, 3	32]	1, 3, 5, 7, 9, 3	33]	1, 3, 6, 3, 3, 3

C. SOCLE DEGREE 5

34]	1, 3, 6, 4, 3, 3	35]	1, 3, 6, 4, 4, 3	36]	1, 3, 6, 4, 5, 3
37]	1, 3, 6, 5, 3, 3	38]	1, 3, 6, 5, 4, 3	39]	1, 3, 6, 5, 5, 3
40]	1, 3, 6, 5, 6, 3	41]	1, 3, 6, 6, 3, 3	42]	1, 3, 6, 6, 4, 3
43]	1, 3, 6, 6, 5, 3	44]	1, 3, 6, 6, 6, 3	45]	1, 3, 6, 6, 7, 3
46]	1, 3, 6, 7, 3, 3	47]	1, 3, 6, 7, 4, 3	48]	1, 3, 6, 7, 5, 3
49]	1, 3, 6, 7, 6, 3	50]	1, 3, 6, 7, 7, 3	51]	1, 3, 6, 7, 8, 3
52]	1, 3, 6, 7, 9, 3	53]	1, 3, 6, 8, 3, 3	54]	1, 3, 6, 8, 4, 3
55]	1, 3, 6, 8, 5, 3	56]	1, 3, 6, 8, 6, 3	57]	1, 3, 6, 8, 7, 3
58]	1, 3, 6, 8, 8, 3	59]	1, 3, 6, 8, 9, 3	60]	1, 3, 6, 9, 3, 3
61]	1, 3, 6, 9, 4, 3	62]	1, 3, 6, 9, 5, 3	63]	1, 3, 6, 9, 6, 3
64]	1, 3, 6, 9, 7, 3	65]	1, 3, 6, 9, 8, 3	66]	1, 3, 6, 9, 9, 3
67]	1, 3, 6, 10, 3, 3	68]	1, 3, 6, 10, 4, 3	69]	1, 3, 6, 10, 5, 3
70]	1, 3, 6, 10, 6, 3	71]	1, 3, 6, 10, 7, 3	72]	1, 3, 6, 10, 8, 3
73]	1, 3, 6, 10, 9, 3				

Non-existence

We first eliminate all sequences we know cannot be the h-vector of a level algebra.

Eliminated by Example 2.13. Using the same sort of reasoning we can eliminate anything of type

$$(1,3,\ldots,\geq 8,4,3), \ (1,3,\ldots,\geq 10,5,3).$$

$(\mathbf{1,3,...,\geq 8,4,3})$: 54], 61], 68].

$(\mathbf{1,3,\ldots,\geq 10,5,3})$: 69].

Eliminated by Corollary 2.11. This corollary easily eliminates 5], 27], 33], 36], and 39].

Eliminated by Corollary 3.5. 2], 3], 4], 6], 8], 9], 14], 16], 17], 20], 21], 26], 34], 37], 40], 41], 46], 52], 53], 60], 67].

Eliminated by Proposition 3.8. Part c): for $d=3$: 15], 35].

Eliminated by Proposition 3.9. 13].

Not cancelable, but not covered by our theorems. 45], 47].

Eliminated by an argument completely analogous to that of Example 3.11. 22], 42], 62].

Eliminated by Example 3.13. 38].

Existence

All the rest of the h-vectors are the h-vector of a level Artinian algebra with the WLP. There are 34 of them and they are:

Table 5.3

1]	1, 3, 3, 3, 3, 3	7]	1, 3, 4, 4, 4, 3	10]	1, 3, 4, 5, 4, 3
11]	1, 3, 4, 5, 5, 3	12]	1, 3, 4, 5, 6, 3	18]	1, 3, 5, 5, 4, 3
19]	1, 3, 5, 5, 5, 3	23]	1, 3, 5, 6, 5, 3	24]	1, 3, 5, 6, 6, 3
25]	1, 3, 5, 6, 7, 3	28]	1, 3, 5, 7, 5, 3	29]	1, 3, 5, 7, 6, 3
30]	1, 3, 5, 7, 7, 3	31]	1, 3, 5, 7, 8, 3	32]	1, 3, 5, 7, 9, 3
43]	1, 3, 6, 6, 5, 3	44]	1, 3, 6, 6, 6, 3	48]	1, 3, 6, 7, 5, 3
49]	1, 3, 6, 7, 6, 3	50]	1, 3, 6, 7, 7, 3	51]	1, 3, 6, 7, 8, 3
55]	1, 3, 6, 8, 5, 3	56]	1, 3, 6, 8, 6, 3	57]	1, 3, 6, 8, 7, 3
58]	1, 3, 6, 8, 8, 3	59]	1, 3, 6, 8, 9, 3	63]	1, 3, 6, 9, 6, 3
64]	1, 3, 6, 9, 7, 3	65]	1, 3, 6, 9, 8, 3	66]	1, 3, 6, 9, 9, 3
70]	1, 3, 6, 10, 6, 3	71]	1, 3, 6, 10, 7, 3	72]	1, 3, 6, 10, 8, 3
73]	1, 3, 6, 10, 9, 3				

We first take advantage of what we already know in socle degree 6.

Using **Theorem 2.2, i)** we get 1], 7], and 10] from, respectively, 5], 25] and 36] in socle degree 6. Thus, these h-vectors are also the h-vectors of level sets of points in $\mathbb{P}^3$ and of Artinian level algebras with the WLP.

We can also apply **Proposition 5.16** to show that 19], 23], 28], 44], 49] and 63] are the h-vectors of level sets of points in $\mathbb{P}^3$ and of Artinian level algebras with the WLP.

We can apply **Proposition 5.24** to get level Artinian algebras with the WLP from **Table 5.2** for the following: 18] from 13], 24] from 19], 48] from 57], 50] from 59], 55] from 62], 56] from 41], 57] from 42], 64] from 69], 70] from 73], 71] from 74].

We will now use the Linked-Sum method to show that all the remaining h-vectors arise as the h-vector of a level algebra. By applying **Proposition 5.15** we see that all the constructed examples also have the WLP. (We continue with the notation employed above.)

11]. $\mathbb{Z} = \left\{ \begin{matrix} \bullet \\ * & \bullet \\ * & \bullet & \bullet \\ * & \bullet & \bullet \\ \bullet & \bullet & \bullet \\ * & \bullet & * \\ \bullet & \bullet & \bullet \end{matrix} \right\}$
$\mathbf{H}(\mathbb{Z}, -)$: 1 3 6 9 12 15 18 $\to$
$\mathbf{H}(\mathbb{X}, -)$: 1 3 6 9 12 13 13 $\to$
$\mathbf{H}(\mathbb{Y}, -)$: 1 3 4 5 5 5 5 $\to$
$\mathbf{H}(A, -)$: 1 3 4 5 5 3 0 $\to$.

12]. $\mathbb{Z} = \left\{ \begin{matrix} * \\ * \\ * & \bullet & \bullet \\ * & \bullet & \bullet & \bullet \\ \bullet & \bullet & \bullet & \bullet \\ * & \bullet & \bullet & * \\ \bullet & \bullet & \bullet & \bullet \end{matrix} \right\}$
$\mathbf{H}(\mathbb{Z}, -)$: 1 3 6 10 14 18 21 $\to$
$\mathbf{H}(\mathbb{X}, -)$: 1 3 6 10 14 15 15 $\to$
$\mathbf{H}(\mathbb{Y}, -)$: 1 3 4 5 6 6 6 $\to$
$\mathbf{H}(A, -)$: 1 3 4 5 6 3 0 $\to$.

C. SOCLE DEGREE 5

25]. $\mathbb{Z} = \left\{\begin{matrix} \bullet & & & \\ * & & & \\ \bullet & \bullet & \bullet & \\ * & \bullet & * & \bullet \\ * & \bullet & \bullet & \bullet \\ * & * & \bullet & \bullet \\ * & \bullet & \bullet & \bullet \end{matrix}\right\}$
$\quad\quad\begin{array}{llllllllll}
\mathbf{H}(\mathbb{Z},-) & : & 1 & 3 & 6 & 10 & 14 & 18 & 21 & \to \\
\mathbf{H}(\mathbb{X},-) & : & 1 & 3 & 6 & 10 & 14 & 14 & 14 & \to \\
\mathbf{H}(\mathbb{Y},-) & : & 1 & 3 & 5 & 6 & 7 & 7 & 7 & \to \\
\mathbf{H}(A,-) & : & 1 & 3 & 5 & 6 & 7 & 3 & 0 & \to.
\end{array}$

29]. $\mathbb{Z} = \left\{\begin{matrix} * & & & \\ * & & & \\ \bullet & \bullet & \bullet & \\ \bullet & \bullet & \bullet & * \\ * & \bullet & \bullet & \bullet \\ * & \bullet & \bullet & * \\ \bullet & \bullet & \bullet & * \end{matrix}\right\}$
$\quad\quad\begin{array}{llllllllll}
\mathbf{H}(\mathbb{Z},-) & : & 1 & 3 & 6 & 10 & 14 & 18 & 21 & \to \\
\mathbf{H}(\mathbb{X},-) & : & 1 & 3 & 6 & 10 & 13 & 14 & 14 & \to \\
\mathbf{H}(\mathbb{Y},-) & : & 1 & 3 & 5 & 7 & 7 & 7 & 7 & \to \\
\mathbf{H}(A,-) & : & 1 & 3 & 5 & 7 & 6 & 3 & 0 & \to.
\end{array}$

30]. $\mathbb{Z} = \left\{\begin{matrix} * & & & \\ \bullet & & & \\ * & * & & \\ \bullet & \bullet & \bullet & \bullet \\ * & * & \bullet & \bullet \\ \bullet & \bullet & \bullet & \bullet \\ * & * & \bullet & \bullet \end{matrix}\right\}$
$\quad\quad\begin{array}{llllllllll}
\mathbf{H}(\mathbb{Z},-) & : & 1 & 3 & 6 & 10 & 14 & 18 & 21 & \to \\
\mathbf{H}(\mathbb{X},-) & : & 1 & 3 & 6 & 10 & 14 & 14 & 14 & \to \\
\mathbf{H}(\mathbb{Y},-) & : & 1 & 3 & 5 & 7 & 7 & 7 & 7 & \to \\
\mathbf{H}(A,-) & : & 1 & 3 & 5 & 7 & 7 & 3 & 0 & \to.
\end{array}$

31]. $\mathbb{Z} = \left\{\begin{matrix} \bullet & & & & \\ \bullet & & & & \\ \bullet & * & * & & \\ * & \bullet & \bullet & & \\ * & \bullet & \bullet & \bullet & * \\ \bullet & \bullet & \bullet & \bullet & * \\ \bullet & \bullet & * & * & \bullet \end{matrix}\right\}$
$\quad\quad\begin{array}{llllllllll}
\mathbf{H}(\mathbb{Z},-) & : & 1 & 3 & 6 & 10 & 15 & 20 & 23 & \to \\
\mathbf{H}(\mathbb{X},-) & : & 1 & 3 & 6 & 10 & 15 & 15 & 15 & \to \\
\mathbf{H}(\mathbb{Y},-) & : & 1 & 3 & 5 & 7 & 8 & 8 & 8 & \to \\
\mathbf{H}(A,-) & : & 1 & 3 & 5 & 7 & 8 & 3 & 0 & \to.
\end{array}$

32]. $\mathbb{Z} = \left\{\begin{matrix} * & * & & & & \\ \bullet & * & & & & \\ \bullet & * & * & \bullet & & \\ \bullet & \bullet & \bullet & * & & \\ \bullet & * & \bullet & \bullet & * & \bullet \\ \bullet & * & \bullet & \bullet & \bullet & \bullet \end{matrix}\right\}$
$\quad\quad\begin{array}{llllllllll}
\mathbf{H}(\mathbb{Z},-) & : & 1 & 3 & 6 & 10 & 15 & 21 & 24 & \to \\
\mathbf{H}(\mathbb{X},-) & : & 1 & 3 & 6 & 10 & 15 & 15 & 15 & \to \\
\mathbf{H}(\mathbb{Y},-) & : & 1 & 3 & 5 & 7 & 9 & 9 & 9 & \to \\
\mathbf{H}(A,-) & : & 1 & 3 & 5 & 7 & 9 & 3 & 0 & \to.
\end{array}$

43]. $\mathbb{Z} = \left\{\begin{matrix} \bullet & & & \\ \bullet & & & \\ \bullet & * & \bullet & \\ \bullet & \bullet & * & \bullet \\ \bullet & \bullet & * & \bullet \\ * & * & \bullet & * \\ \bullet & \bullet & \bullet & \bullet \end{matrix}\right\}$
$\quad\quad\begin{array}{llllllllll}
\mathbf{H}(\mathbb{Z},-) & : & 1 & 3 & 6 & 10 & 14 & 18 & 21 & \to \\
\mathbf{H}(\mathbb{X},-) & : & 1 & 3 & 6 & 10 & 13 & 15 & 15 & \to \\
\mathbf{H}(\mathbb{Y},-) & : & 1 & 3 & 6 & 6 & 6 & 6 & 6 & \to \\
\mathbf{H}(A,-) & : & 1 & 3 & 6 & 6 & 5 & 3 & 0 & \to.
\end{array}$

51]. $\mathbb{Z} = \left\{ \begin{matrix} \bullet \\ * \\ * & \bullet & \bullet \\ \bullet & * & \bullet \\ \bullet & \bullet & * & \bullet & * \\ \bullet & \bullet & * & * & \bullet \\ \bullet & \bullet & \bullet & \bullet & * \end{matrix} \right\}$
$\begin{array}{lccccccccc} \mathbf{H}(\mathbb{Z},-) & : & 1 & 3 & 6 & 10 & 15 & 20 & 23 & \to \\ \mathbf{H}(\mathbb{X},-) & : & 1 & 3 & 6 & 10 & 15 & 15 & 15 & \to \\ \mathbf{H}(\mathbb{Y},-) & : & 1 & 3 & 6 & 7 & 8 & 8 & 8 & \to \\ \mathbf{H}(A,-) & : & 1 & 3 & 6 & 7 & 8 & 3 & 0 & \to. \end{array}$

58]. $\mathbb{Z} = \left\{ \begin{matrix} * \\ * \\ \bullet & \bullet & \bullet \\ \bullet & \bullet & \bullet \\ \bullet & * & \bullet & * \\ \bullet & * & * & * & \bullet \\ \bullet & * & \bullet & \bullet & \bullet \end{matrix} \right\}$
$\begin{array}{lccccccccc} \mathbf{H}(\mathbb{Z},-) & : & 1 & 3 & 6 & 10 & 15 & 20 & 23 & \to \\ \mathbf{H}(\mathbb{X},-) & : & 1 & 3 & 6 & 10 & 15 & 15 & 15 & \to \\ \mathbf{H}(\mathbb{Y},-) & : & 1 & 3 & 6 & 8 & 8 & 8 & 8 & \to \\ \mathbf{H}(A,-) & : & 1 & 3 & 6 & 8 & 8 & 3 & 0 & \to. \end{array}$

59]. $\mathbb{Z} = \left\{ \begin{matrix} \bullet & * \\ \bullet & * \\ \bullet & \bullet & \bullet & \bullet \\ * & \bullet & \bullet & \bullet \\ * & * & * & \bullet & * & * \\ \bullet & \bullet & \bullet & * & \bullet & \bullet \end{matrix} \right\}$
$\begin{array}{lccccccccc} \mathbf{H}(\mathbb{Z},-) & : & 1 & 3 & 6 & 10 & 15 & 21 & 24 & \to \\ \mathbf{H}(\mathbb{X},-) & : & 1 & 3 & 6 & 10 & 15 & 15 & 15 & \to \\ \mathbf{H}(\mathbb{Y},-) & : & 1 & 3 & 6 & 8 & 9 & 9 & 9 & \to \\ \mathbf{H}(A,-) & : & 1 & 3 & 6 & 8 & 9 & 3 & 0 & \to. \end{array}$

65]. $\mathbb{Z} = \left\{ \begin{matrix} \bullet & * \\ * & \bullet \\ \bullet & * & \bullet & * \\ \bullet & \bullet & \bullet & * \\ * & \bullet & \bullet & \bullet & \bullet & * \\ * & * & \bullet & \bullet & \bullet & \bullet \end{matrix} \right\}$
$\begin{array}{lccccccccc} \mathbf{H}(\mathbb{Z},-) & : & 1 & 3 & 6 & 10 & 15 & 21 & 24 & \to \\ \mathbf{H}(\mathbb{X},-) & : & 1 & 3 & 6 & 10 & 14 & 15 & 15 & \to \\ \mathbf{H}(\mathbb{Y},-) & : & 1 & 3 & 6 & 9 & 9 & 9 & 9 & \to \\ \mathbf{H}(A,-) & : & 1 & 3 & 6 & 9 & 8 & 3 & 0 & \to. \end{array}$

66]. $\mathbb{Z} = \left\{ \begin{matrix} \bullet & \bullet \\ * & * \\ * & * & \bullet & * \\ \bullet & \bullet & \bullet & \bullet \\ \bullet & \bullet & \bullet & \bullet & \bullet & * \\ * & \bullet & * & \bullet & \bullet & * \end{matrix} \right\}$
$\begin{array}{lccccccccc} \mathbf{H}(\mathbb{Z},-) & : & 1 & 3 & 6 & 10 & 15 & 21 & 24 & \to \\ \mathbf{H}(\mathbb{X},-) & : & 1 & 3 & 6 & 10 & 15 & 15 & 15 & \to \\ \mathbf{H}(\mathbb{Y},-) & : & 1 & 3 & 6 & 9 & 9 & 9 & 9 & \to \\ \mathbf{H}(A,-) & : & 1 & 3 & 6 & 9 & 9 & 3 & 0 & \to. \end{array}$

72]. $\mathbb{Z} = \left\{ \begin{matrix} * \\ \bullet \\ \bullet & * & * \\ \bullet & * & \bullet \\ * & \bullet & \bullet & * & \bullet \\ * & * & * & \bullet & \bullet \\ \bullet & \bullet & * & \bullet & \bullet \end{matrix} \right\}$
$\begin{array}{lccccccccc} \mathbf{H}(\mathbb{Z},-) & : & 1 & 3 & 6 & 10 & 15 & 20 & 23 & \to \\ \mathbf{H}(\mathbb{X},-) & : & 1 & 3 & 6 & 10 & 13 & 13 & 13 & \to \\ \mathbf{H}(\mathbb{Y},-) & : & 1 & 3 & 6 & 10 & 10 & 10 & 10 & \to \\ \mathbf{H}(A,-) & : & 1 & 3 & 6 & 10 & 8 & 3 & 0 & \to. \end{array}$

73]. $\mathbb{Z} = \left\{ \begin{matrix} \bullet & \bullet & & & & \\ * & * & & & & \\ \bullet & \bullet & * & \bullet & & \\ * & * & \bullet & * & & \\ \bullet & * & * & \bullet & \bullet & \bullet \\ * & \bullet & \bullet & * & \bullet & \bullet \end{matrix} \right\}$
$\begin{array}{llllllllll} \mathbf{H}(\mathbb{Z},-) & : & 1 & 3 & 6 & 10 & 15 & 21 & 24 & \to \\ \mathbf{H}(\mathbb{X},-) & : & 1 & 3 & 6 & 10 & 14 & 14 & 14 & \to \\ \mathbf{H}(\mathbb{Y},-) & : & 1 & 3 & 6 & 10 & 10 & 10 & 10 & \to \\ \mathbf{H}(A,-) & : & 1 & 3 & 6 & 10 & 9 & 3 & 0 & \to . \end{array}$

Socle Degree 5 and Type 4

After one eliminates a few trivial cases one is left with a list of 62 tuples which could possibly be the h-vector of an Artinian level algebra of socle degree 5 and type 4. They are in the list below.

Table 5.4A

1]	1, 3, 3, 4, 4, 4	2]	1, 3, 3, 4, 5, 4	3]	1, 3, 4, 4, 4, 4
4]	1, 3, 4, 4, 5, 4	5]	1, 3, 4, 5, 4, 4	6]	1, 3, 4, 5, 5, 4
7]	1, 3, 4, 5, 6, 4	8]	1, 3, 5, 4, 4, 4	9]	1, 3, 5, 4, 5, 4
10]	1, 3, 5, 5, 4, 4	11]	1, 3, 5, 5, 5, 4	12]	1, 3, 5, 5, 6, 4
13]	1, 3, 5, 6, 4, 4	14]	1, 3, 5, 6, 5, 4	15]	1, 3, 5, 6, 6, 4
16]	1, 3, 5, 6, 7, 4	17]	1, 3, 5, 7, 4, 4	18]	1, 3, 5, 7, 5, 4
19]	1, 3, 5, 7, 6, 4	20]	1, 3, 5, 7, 7, 4	21]	1, 3, 5, 7, 8, 4
22]	1, 3, 5, 7, 9, 4	23]	1, 3, 6, 4, 4, 4	24]	1, 3, 6, 4, 5, 4
25]	1, 3, 6, 5, 4, 4	26]	1, 3, 6, 5, 5, 4	27]	1, 3, 6, 5, 6, 4
28]	1, 3, 6, 6, 4, 4	29]	1, 3, 6, 6, 5, 4	30]	1, 3, 6, 6, 6, 4
31]	1, 3, 6, 6, 7, 4	32]	1, 3, 6, 7, 4, 4	33]	1, 3, 6, 7, 5, 4
34]	1, 3, 6, 7, 6, 4	35]	1, 3, 6, 7, 7, 4	36]	1, 3, 6, 7, 8, 4
37]	1, 3, 6, 7, 9, 4	38]	1, 3, 6, 8, 4, 4	39]	1, 3, 6, 8, 5, 4
40]	1, 3, 6, 8, 6, 4	41]	1, 3, 6, 8, 7, 4	42]	1, 3, 6, 8, 8, 4
43]	1, 3, 6, 8, 9, 4	44]	1, 3, 6, 8, 10, 4	45]	1, 3, 6, 9, 4, 4
46]	1, 3, 6, 9, 5, 4	47]	1, 3, 6, 9, 6, 4	48]	1, 3, 6, 9, 7, 4
49]	1, 3, 6, 9, 8, 4	50]	1, 3, 6, 9, 9, 4	51]	1, 3, 6, 9, 10, 4
52]	1, 3, 6, 9, 11, 4	53]	1, 3, 6, 9, 12, 4	54]	1, 3, 6, 10, 4, 4
55]	1, 3, 6, 10, 5, 4	56]	1, 3, 6, 10, 6, 4	57]	1, 3, 6, 10, 7, 4
58]	1, 3, 6, 10, 8, 4	59]	1, 3, 6, 10, 9, 4	60]	1, 3, 6, 10, 10, 4
61]	1, 3, 6, 10, 11, 4	62]	1, 3, 6, 10, 12, 4		

Non-existence

Eliminate by Corollary 3.5. 1], 2], 4], 9], 12], 24], 27], and 37].

Eliminated by Proposition 3.8.
Part b): for $h_4 = 4 = h_5$ and $h_3 > 4$: 5], 10], 13], 17], 25], 28], 32], 38], 45], and 54].
Part c): for $h_3 = h_4 = 4$ and $h_2 > 4$: 8] and 23].

By arguments very similar to those of Example 3.11, we can eliminate 18], 26], 33], and 56].

The following are not cancelable, but this is not a consequence of any of our theorems 31], 39], 46], and 55].

Existence

This leaves the following 34 possibilities.

Table 5.4

3]	1, 3, 4, 4, 4, 4	6]	1, 3, 4, 5, 5, 4	7]	1, 3, 4, 5, 6, 4
11]	1, 3, 5, 5, 5, 4	14]	1, 3, 5, 6, 5, 4	15]	1, 3, 5, 6, 6, 4
16]	1, 3, 5, 6, 7, 4	19]	1, 3, 5, 7, 6, 4	20]	1, 3, 5, 7, 7, 4
21]	1, 3, 5, 7, 8, 4	22]	1, 3, 5, 7, 9, 4	29]	1, 3, 6, 6, 5, 4
30]	1, 3, 6, 6, 6, 4	34]	1, 3, 6, 7, 6, 4	35]	1, 3, 6, 7, 7, 4
36]	1, 3, 6, 7, 8, 4	40]	1, 3, 6, 8, 6, 4	41]	1, 3, 6, 8, 7, 4
42]	1, 3, 6, 8, 8, 4	43]	1, 3, 6, 8, 9, 4	44]	1, 3, 6, 8, 10, 4
47]	1, 3, 6, 9, 6, 4	48]	1, 3, 6, 9, 7, 4	49]	1, 3, 6, 9, 8, 4
50]	1, 3, 6, 9, 9, 4	51]	1, 3, 6, 9, 10, 4	52]	1, 3, 6, 9, 11, 4
53]	1, 3, 6, 9, 12, 4	57]	1, 3, 6, 10, 7, 4	58]	1, 3, 6, 10, 8, 4
59]	1, 3, 6, 10, 9, 4	60]	1, 3, 6, 10, 10, 4	61]	1, 3, 6, 10, 11, 4
62]	1, 3, 6, 10, 12, 4				

The following come from examples we made for socle degree 6 and type 2 (by truncation) and so are also the h-vectors of Artinian level algebras with the WLP: 3], 6], 11], 14], 15], 19], 30], 34], 35], 40], 41], 48], and 57].

We can apply **Proposition 5.24** to get level Artinian algebras with the WLP from **Table 5.3** for the following: 16] from 12], 29] from 18], 36] from 25], 42] from 50], 43] from 51], 44] from 32], 47] from 55], 49] from 57], 50] from 58], 51] from 59], 58] from 64], 59] from 65], 60] from 66].

52]. We get a level Artinian algebra with the WLP from **Example 6.18**.

The following can be constructed with the linked-sum method. All those constructed with the linked-sum method can also be seen (via **Proposition 5.15**) to satisfy the WLP. (The notation for the linked-sum constructions is as above.)

7]. $\mathbb{Z} = \left\{ \begin{matrix} * \\ * & \bullet \\ * & \bullet & \bullet \\ \bullet & \bullet & \bullet & \bullet \\ * & \bullet & * & \bullet \\ * & \bullet & \bullet & \bullet \\ \bullet & \bullet & \bullet & \bullet \end{matrix} \right\}$

$\mathbf{H}(\mathbb{Z}, -) : 1 \quad 3 \quad 6 \quad 10 \quad 14 \quad 18 \quad 22 \quad \to$
$\mathbf{H}(\mathbb{X}, -) : 1 \quad 3 \quad 6 \quad 10 \quad 14 \quad 16 \quad 16 \quad \to$
$\mathbf{H}(\mathbb{Y}, -) : 1 \quad 3 \quad 4 \quad 5 \quad 6 \quad 6 \quad 6 \quad \to$
$\mathbf{H}(A, -) : 1 \quad 3 \quad 4 \quad 5 \quad 6 \quad 4 \quad 0 \quad \to .$

20]. $\mathbb{Z} = \left\{ \begin{matrix} \bullet \\ * & * \\ * & * & \bullet \\ \bullet & \bullet & \bullet & \bullet \\ * & \bullet & \bullet & \bullet \\ * & * & \bullet & \bullet \\ \bullet & \bullet & \bullet & \bullet \end{matrix} \right\}$

$\mathbf{H}(\mathbb{Z}, -) : 1 \quad 3 \quad 6 \quad 10 \quad 14 \quad 18 \quad 22 \quad \to$
$\mathbf{H}(\mathbb{X}, -) : 1 \quad 3 \quad 6 \quad 10 \quad 14 \quad 15 \quad 15 \quad \to$
$\mathbf{H}(\mathbb{Y}, -) : 1 \quad 3 \quad 5 \quad 7 \quad 7 \quad 7 \quad 7 \quad \to$
$\mathbf{H}(A, -) : 1 \quad 3 \quad 5 \quad 7 \quad 7 \quad 4 \quad 0 \quad \to .$

21].
$$\mathbb{Z} = \left\{ \begin{matrix} \bullet & & & \\ * & \bullet & & \\ * & \bullet & \bullet & \\ \bullet & * & * & * \\ * & \bullet & \bullet & \bullet \\ * & \bullet & \bullet & \bullet \\ * & \bullet & \bullet & \bullet \end{matrix} \right\}$$

$\mathbf{H}(\mathbb{Z}, -)$: 1 3 6 10 14 18 22 $\to$
$\mathbf{H}(\mathbb{X}, -)$: 1 3 6 10 14 14 14 $\to$
$\mathbf{H}(\mathbb{Y}, -)$: 1 3 5 7 8 8 8 $\to$
$\mathbf{H}(A, -)$: 1 3 5 7 8 4 0 $\to$.

22].
$$\mathbb{Z} = \left\{ \begin{matrix} \bullet & * & & & & & \\ * & * & & & & & \\ \bullet & * & \bullet & \bullet & & & \\ \bullet & * & * & \bullet & & & \\ \bullet & * & \bullet & * & \bullet & \bullet & \\ \bullet & \bullet & \bullet & \bullet & * & \bullet & \bullet \end{matrix} \right\}$$

$\mathbf{H}(\mathbb{Z}, -)$: 1 3 6 10 15 21 25 $\to$
$\mathbf{H}(\mathbb{X}, -)$: 1 3 6 10 15 16 16 $\to$
$\mathbf{H}(\mathbb{Y}, -)$: 1 3 5 7 9 9 9 $\to$
$\mathbf{H}(A, -)$: 1 3 5 7 9 4 0 $\to$.

The cases still remaining are 53], 61], and 62]. None of these can be constructed using the linked-sum method (see **Remark 5.32**). We use some results of Iarrobino [**43**] to construct 62] and 61].

53]. This is the h-vector of 35 general points on a smooth cubic surface in $\mathbb{P}^3$.

61]. This can be constructed using an Inverse System generated by the following 4 forms of degree 5:

$$F_1 = \sum_{i=1}^{3} L_i^5 \quad F_2 = \sum_{i=1}^{3} M_i^5 \quad F_3 = \sum_{i=1}^{3} N_i^5 \quad F_4 = H_1^5 + H_2^5$$

where the L_i, M_i, N_i and H_i are generic linear forms.

62]. This can be constructed using the Inverse System generated by 4 generic forms of degree 5. This is a compressed level algebra.

It remains to find level examples with the WLP for 53], 61] and 62].

53]. One shows that the 35 points on a smooth cubic surface (mentioned above) actually has the property that the Artinian reduction of its coordinate ring also satisfies the WLP.

61] and 62]. As in **Example 6.18** we could find (using a computer) an example (in each case) of a set of points on an ACM curve whose homogeneous coordinate ring had the property that its Artinian reduction was level and had the WLP.

Socle Degree 5 and Type 5

In this case there are 51 possible tuples which could be the h-vector of such a level algebra. We begin by eliminating those we can.

Table 5.5A

1]	1, 3, 3, 4, 5, 5	2]	1, 3, 4, 4, 5, 5	3]	1, 3, 4, 5, 5, 5
4]	1, 3, 4, 5, 6, 5	5]	1, 3, 5, 4, 5, 5	6]	1, 3, 5, 5, 5, 5
7]	1, 3, 5, 5, 6, 5	8]	1, 3, 5, 6, 5, 5	9]	1, 3, 5, 6, 6, 5
10]	1, 3, 5, 6, 7, 5	11]	1, 3, 5, 7, 5, 5	12]	1, 3, 5, 7, 6, 5
13]	1, 3, 5, 7, 7, 5	14]	1, 3, 5, 7, 8, 5	15]	1, 3, 5, 7, 9, 5
16]	1, 3, 6, 4, 5, 5	17]	1, 3, 6, 5, 5, 5	18]	1, 3, 6, 5, 6, 5
19]	1, 3, 6, 6, 5, 5	20]	1, 3, 6, 6, 6, 5	21]	1, 3, 6, 6, 7, 5
22]	1, 3, 6, 7, 5, 5	23]	1, 3, 6, 7, 6, 5	24]	1, 3, 6, 7, 7, 5
25]	1, 3, 6, 7, 8, 5	26]	1, 3, 6, 7, 9, 5	27]	1, 3, 6, 8, 5, 5
28]	1, 3, 6, 8, 6, 5	29]	1, 3, 6, 8, 7, 5	30]	1, 3, 6, 8, 8, 5
31]	1, 3, 6, 8, 9, 5	32]	1, 3, 6, 8, 10, 5	33]	1, 3, 6, 9, 5, 5
34]	1, 3, 6, 9, 6, 5	35]	1, 3, 6, 9, 7, 5	36]	1, 3, 6, 9, 8, 5
37]	1, 3, 6, 9, 9, 5	38]	1, 3, 6, 9, 10, 5	39]	1, 3, 6, 9, 11, 5
40]	1, 3, 6, 9, 12, 5	41]	1, 3, 6, 10, 5, 5	42]	1, 3, 6, 10, 6, 5
43]	1, 3, 6, 10, 7, 5	44]	1, 3, 6, 10, 8, 5	45]	1, 3, 6, 10, 9, 5
46]	1, 3, 6, 10, 10, 5	47]	1, 3, 6, 10, 11, 5	48]	1, 3, 6, 10, 12, 5
49]	1, 3, 6, 10, 13, 5	50]	1, 3, 6, 10, 14, 5	51]	1, 3, 6, 10, 15, 5

Non-existence

Eliminated by Corollary 3.5. 1], 2], 5], 7], 16], 18], 26].

Eliminated by Proposition 3.8, c). $h_4 = h_5 = 5$ and $h_3 > 5$: 8], 11], 19], 22], 27], 33], 41].

Eliminated by an argument completely analogous to that of Example 3.11. 17], [28].

The following are not cancelable, but not from any of our theorems. 21], 34], 42].

Existence

This leaves the following as possible h-vectors of level algebras.

Table 5.5

3]	1, 3, 4, 5, 5, 5	4]	1, 3, 4, 5, 6, 5	6]	1, 3, 5, 5, 5, 5
9]	1, 3, 5, 6, 6, 5	10]	1, 3, 5, 6, 7, 5	12]	1, 3, 5, 7, 6, 5
13]	1, 3, 5, 7, 7, 5	14]	1, 3, 5, 7, 8, 5	15]	1, 3, 5, 7, 9, 5
20]	1, 3, 6, 6, 6, 5	23]	1, 3, 6, 7, 6, 5	24]	1, 3, 6, 7, 7, 5
25]	1, 3, 6, 7, 8, 5	29]	1, 3, 6, 8, 7, 5	30]	1, 3, 6, 8, 8, 5
31]	1, 3, 6, 8, 9, 5	32]	1, 3, 6, 8, 10, 5	35]	1, 3, 6, 9, 7, 5
36]	1, 3, 6, 9, 8, 5	37]	1, 3, 6, 9, 9, 5	38]	1, 3, 6, 9, 10, 5
39]	1, 3, 6, 9, 11, 5	40]	1, 3, 6, 9, 12, 5	43]	1, 3, 6, 10, 7, 5
44]	1, 3, 6, 10, 8, 5	45]	1, 3, 6, 10, 9, 5	46]	1, 3, 6, 10, 10, 5
47]	1, 3, 6, 10, 11, 5	48]	1, 3, 6, 10, 12, 5	49]	1, 3, 6, 10, 13, 5
50]	1, 3, 6, 10, 14, 5	51]	1, 3, 6, 10, 15, 5		

Of these, the following come from socle degree 6 and type 2 (by truncation) and so are also the h-vectors of level Artinian algebras with the WLP: 3], 4], 6], 9], 10], 12], 13], 14], 20], 23], 24], 25], 29], 30], 31], 35], 36], 37], 44], 45].

C. SOCLE DEGREE 5

We can apply **Proposition 5.24** to get level Artinian algebras with the WLP from **Table 5.4** for the following: 32] from 22], 38] from 43], 39] from 44], 43] from 47], 46] from 50], 47] from 51], 48] from 52], 49] from 62].

51] is the h-vector of a compressed level algebra. By **Proposition 5.22** any level algebra with this h-vector has the WLP.

15] can be constructed using the linked-sum method. It is easily seen to satisfy the WLP (**Proposition 5.15**).

15]. $\mathbb{Z} = \left\{\begin{array}{cccccc} * & & & & & \\ * & \bullet & & & & \\ * & * & \bullet & & & \\ \bullet & \bullet & \bullet & \bullet & & \\ * & * & \bullet & \bullet & \bullet & \\ * & * & \bullet & \bullet & \bullet & \\ \bullet & * & \bullet & \bullet & \bullet & \end{array}\right\}$
$\mathbf{H}(\mathbb{Z},-)$: 1 3 6 10 15 20 25 $\rightarrow$
$\mathbf{H}(\mathbb{X},-)$: 1 3 6 10 15 16 16 $\rightarrow$
$\mathbf{H}(\mathbb{Y},-)$: 1 3 5 7 9 9 9 $\rightarrow$
$\mathbf{H}(A,-)$: 1 3 5 7 9 5 0 $\rightarrow$.

The cases still remaining are: 40] and 50]. Neither can be constructed using the linked-sum method. We now construct examples for each of these cases by other means.

40]. This is the h-vector of 36 points on a smooth cubic surface in $\mathbb{P}^3$.

50]. We can construct this using Inverse Systems by choosing 5 forms of degree 5 of the type

$$F_1 = \sum_{i=1}^3 L_i^5 \quad F_2 = \sum_{i=1}^3 M_i^5 \quad F_3 = \sum_{i=1}^3 N_i^5 \quad F_4 = \sum_{i=1}^3 H_i^5 \quad F_5 = G_1^5 + G_2^5$$

where the L_i, M_i, N_i, H_i and G_i are generic linear forms.

It remains to find level examples with the WLP for 40] and 50].

40]. As in earlier cases, we can take 36 general points on a smooth cubic surface in $\mathbb{P}^3$, whose homogeneous coordinate ring has Artinian reduction with the WLP.

50]. As in **Example 6.18** we can find (with a computer) a set of 39 points on an sufficiently general ACM curve of degree 14 and genus 23 whose homogeneous coordinate ring has a level Artinian reduction with the WLP.

Socle Degree 5 and Type 6

From the 51 possible h-vectors in **Table 5.6A** we first eliminate those we can.

Table 5.6A

1]	1, 3, 3, 4, 5, 6	2]	1, 3, 4, 4, 5, 6	3]	1, 3, 4, 5, 5, 6
4]	1, 3, 4, 5, 6, 6	5]	1, 3, 5, 4, 5, 6	6]	1, 3, 5, 5, 5, 6
7]	1, 3, 5, 5, 6, 6	8]	1, 3, 5, 6, 5, 6	9]	1, 3, 5, 6, 6, 6
10]	1, 3, 5, 6, 7, 6	11]	1, 3, 5, 7, 5, 6	12]	1, 3, 5, 7, 6, 6
13]	1, 3, 5, 7, 7, 6	14]	1, 3, 5, 7, 8, 6	15]	1, 3, 5, 7, 9, 6
16]	1, 3, 6, 4, 5, 6	17]	1, 3, 6, 5, 5, 6	18]	1, 3, 6, 5, 6, 6
19]	1, 3, 6, 6, 5, 6	20]	1, 3, 6, 6, 6, 6	21]	1, 3, 6, 6, 7, 6
22]	1, 3, 6, 7, 5, 6	23]	1, 3, 6, 7, 6, 6	24]	1, 3, 6, 7, 7, 6
25]	1, 3, 6, 7, 8, 6	26]	1, 3, 6, 7, 9, 6	27]	1, 3, 6, 8, 5, 6
28]	1, 3, 6, 8, 6, 6	29]	1, 3, 6, 8, 7, 6	30]	1, 3, 6, 8, 8, 6
31]	1, 3, 6, 8, 9, 6	32]	1, 3, 6, 8, 10, 6	33]	1, 3, 6, 9, 5, 6
34]	1, 3, 6, 9, 6, 6	35]	1, 3, 6, 9, 7, 6	36]	1, 3, 6, 9, 8, 6
37]	1, 3, 6, 9, 9, 6	38]	1, 3, 6, 9, 10, 6	39]	1, 3, 6, 9, 11, 6
40]	1, 3, 6, 9, 12, 6	41]	1, 3, 6, 10, 5, 6	42]	1, 3, 6, 10, 6, 6
43]	1, 3, 6, 10, 7, 6	44]	1, 3, 6, 10, 8, 6	45]	1, 3, 6, 10, 9, 6
46]	1, 3, 6, 10, 10, 6	47]	1, 3, 6, 10, 11, 6	48]	1, 3, 6, 10, 12, 6
49]	1, 3, 6, 10, 13, 6	50]	1, 3, 6, 10, 14, 6	51]	1, 3, 6, 10, 15, 6

Non-existence

Eliminated by Corollary 3.5. 1], 2], 3], 5], 6], 7], 8], 11], 16], 17], 18], 19], 22], 26], 27], 33], 41].

Eliminated by Proposition 3.9. 28], 34], 42].

Eliminated by an argument completely analogous to that of Example 3.11. 12], 23], 43].

Eliminated by an argument similar to that of Example 3.12. 35].

Not cancelable but not because of any of our theorems. 21].

Existence

This leaves the following as possible h-vectors of level artinian algebras.

Table 5.6

4]	1, 3, 4, 5, 6, 6	9]	1, 3, 5, 6, 6, 6	10]	1, 3, 5, 6, 7, 6
13]	1, 3, 5, 7, 7, 6	14]	1, 3, 5, 7, 8, 6	15]	1, 3, 5, 7, 9, 6
20]	1, 3, 6, 6, 6, 6	24]	1, 3, 6, 7, 7, 6	25]	1, 3, 6, 7, 8, 6
29]	1, 3, 6, 8, 7, 6	30]	1, 3, 6, 8, 8, 6	31]	1, 3, 6, 8, 9, 6
32]	1, 3, 6, 8, 10, 6	36]	1, 3, 6, 9, 8, 6	37]	1, 3, 6, 9, 9, 6
38]	1, 3, 6, 9, 10, 6	39]	1, 3, 6, 9, 11, 6	40]	1, 3, 6, 9, 12, 6
44]	1, 3, 6, 10, 8, 6	45]	1, 3, 6, 10, 9, 6	46]	1, 3, 6, 10, 10, 6
47]	1, 3, 6, 10, 11, 6	48]	1, 3, 6, 10, 12, 6	49]	1, 3, 6, 10, 13, 6
50]	1, 3, 6, 10, 14, 6	51]	1, 3, 6, 10, 15, 6		

Of the h-vectors in the table above we can obtain the following from socle degree 6 and type 2 (by truncation). Consequently these h-vectors correspond to Artinian level algebras with the WLP: 9], 10], 13], 14], 15], 20], 24], 25], 29], 30], 31], 32], 36], 37], 38], 39], 40], 44], 45], 46], 47], 48].

4] corresponds to an Artinian level algebra with the WLP, using **Proposition 5.16**.

49], 50], and 51] come, respectively (using **Proposition 5.24**) from 48], 49], and 51] in **Table 5.5**.

Socle Degree 5 and Type 7

In this case we have 37 possible h-vectors given in **Table 5.7A**.

Table 5.7A

1]	1, 3, 4, 5, 6, 7	2]	1, 3, 5, 5, 6, 7	3]	1, 3, 5, 6, 6, 7
4]	1, 3, 5, 6, 7, 7	5]	1, 3, 5, 7, 6, 7	6]	1, 3, 5, 7, 7, 7
7]	1, 3, 5, 7, 8, 7	8]	1, 3, 5, 7, 9, 7	9]	1, 3, 6, 5, 6, 7
10]	1, 3, 6, 6, 6, 7	11]	1, 3, 6, 6, 7, 7	12]	1, 3, 6, 7, 6, 7
13]	1, 3, 6, 7, 7, 7	14]	1, 3, 6, 7, 8, 7	15]	1, 3, 6, 7, 9, 7
16]	1, 3, 6, 8, 6, 7	17]	1, 3, 6, 8, 7, 7	18]	1, 3, 6, 8, 8, 7
19]	1, 3, 6, 8, 9, 7	20]	1, 3, 6, 8, 10, 7	21]	1, 3, 6, 9, 6, 7
22]	1, 3, 6, 9, 7, 7	23]	1, 3, 6, 9, 8, 7	24]	1, 3, 6, 9, 9, 7
25]	1, 3, 6, 9, 10, 7	26]	1, 3, 6, 9, 11, 7	27]	1, 3, 6, 9, 12, 7
28]	1, 3, 6, 10, 6, 7	29]	1, 3, 6, 10, 7, 7	30]	1, 3, 6, 10, 8, 7
31]	1, 3, 6, 10, 9, 7	32]	1, 3, 6, 10, 10, 7	33]	1, 3, 6, 10, 11, 7
34]	1, 3, 6, 10, 12, 7	35]	1, 3, 6, 10, 13, 7	36]	1, 3, 6, 10, 14, 7
37]	1, 3, 6, 10, 15, 7				

Non-existence

Eliminated by Corollary 3.5. 2], 3], 5], 9], 10], 12], 15], 16], 21], 28].

Eliminated by Proposition 3.9. 22], 29].

Eliminated by methods analogous to Example 3.11. 17].

Eliminated since not cancelable, but not by any of our theorems. 11].

Existence

This leaves the following as possible h-vectors of level Artinian algebras.

Table 5.7

1]	1, 3, 4, 5, 6, 7	4]	1, 3, 5, 6, 7, 7	6]	1, 3, 5, 7, 7, 7
7]	1, 3, 5, 7, 8, 7	8]	1, 3, 5, 7, 9, 7	13]	1, 3, 6, 7, 7, 7
14]	1, 3, 6, 7, 8, 7	18]	1, 3, 6, 8, 8, 7	19]	1, 3, 6, 8, 9, 7
20]	1, 3, 6, 8, 10, 7	23]	1, 3, 6, 9, 8, 7	24]	1, 3, 6, 9, 9, 7
25]	1, 3, 6, 9, 10, 7	26]	1, 3, 6, 9, 11, 7	27]	1, 3, 6, 9, 12, 7
30]	1, 3, 6, 10, 8, 7	31]	1, 3, 6, 10, 9, 7	32]	1, 3, 6, 10, 10, 7
33]	1, 3, 6, 10, 11, 7	34]	1, 3, 6, 10, 12, 7	35]	1, 3, 6, 10, 13, 7
36]	1, 3, 6, 10, 14, 7	37]	1, 3, 6, 10, 15, 7		

Using **Proposition 5.16** we find examples satisfying the WLP for: 1], 4], 6], 7], 8], 13], and 14].

We can apply **Proposition 5.24** to get level Artinian algebras with the WLP from **Table 5.6** for the following: 18] from 24], 19] from 25], 20] from 15], 23] from 29], 24] from 30], 25] from 31], 26] from 32], 31] from 36], 32] from 45], 33] from 46], 34] from 47], 35] from 48], 36] from 49], 37] from 51]).

27] can be constructed by the linked-sum method. From **Proposition 5.15** we see it also satisfies the WLP.

$$\mathbb{Z} = \left\{ \begin{matrix} \bullet & & & & & \\ * & * & & & & \\ * & \bullet & * & & & \\ \bullet & \bullet & \bullet & \bullet & & \\ \bullet & * & \bullet & * & \bullet & \\ \bullet & \bullet & * & \bullet & \bullet & \bullet \\ * & * & \bullet & * & \bullet & * & * \end{matrix} \right\}$$

$\mathbf{H}(\mathbb{Z}, -)$: 1 3 6 10 15 21 28 →
$\mathbf{H}(\mathbb{X}, -)$: 1 3 6 10 15 16 16 →
$\mathbf{H}(\mathbb{Y}, -)$: 1 3 6 9 12 12 12 →
$\mathbf{H}(A, -)$: 1 3 6 9 12 7 0 →.

30] can be constructed as follows. Let J be an Artinian Gorenstein codimension three ideal with Hilbert function $(1, 3, 6, 3, 1)$. Let L be a general linear form and let A be the vector space of all forms of degree 6. Then $I = LJ + A$ can be checked to give a level algebra with the desired Hilbert function. This ideal does not have WLP, though, and we do not know if such an ideal exists with WLP.

Socle Degree 5 and Type 8

There are 27 possible h-vectors we have to consider. They are in the table below:

Table 5.8A

1]	1, 3, 5, 6, 7, 8	2]	1, 3, 5, 7, 7, 8	3]	1, 3, 5, 7, 8, 8
4]	1, 3, 5, 7, 9, 8	5]	1, 3, 6, 6, 7, 8	6]	1, 3, 6, 7, 7, 8
7]	1, 3, 6, 7, 8, 8	8]	1, 3, 6, 7, 9, 8	9]	1, 3, 6, 8, 7, 8
10]	1, 3, 6, 8, 8, 8	11]	1, 3, 6, 8, 9, 8	12]	1, 3, 6, 8, 10, 8
13]	1, 3, 6, 9, 7, 8	14]	1, 3, 6, 9, 8, 8	15]	1, 3, 6, 9, 9, 8
16]	1, 3, 6, 9, 10, 8	17]	1, 3, 6, 9, 11, 8	18]	1, 3, 6, 9, 12, 8
19]	1, 3, 6, 10, 7, 8	20]	1, 3, 6, 10, 8, 8	21]	1, 3, 6, 10, 9, 8
22]	1, 3, 6, 10, 10, 8	23]	1, 3, 6, 10, 11, 8	24]	1, 3, 6, 10, 12, 8
25]	1, 3, 6, 10, 13, 8	26]	1, 3, 6, 10, 14, 8	27]	1, 3, 6, 10, 15, 8

Non-existence

Eliminated using Corollary 3.5. 2], 6], 8], 9], 13], 19].

Eliminated using an argument completely analogous to Example 3.11. 14].

These are not cancelable, but not because of any of our theorems. 5], 20].

Existence

This leaves the following as possible h-vectors of level algebras:

Table 5.8

1]	1, 3, 5, 6, 7, 8	3]	1, 3, 5, 7, 8, 8	4]	1, 3, 5, 7, 9, 8
7]	1, 3, 6, 7, 8, 8	10]	1, 3, 6, 8, 8, 8	11]	1, 3, 6, 8, 9, 8
12]	1, 3, 6, 8, 10, 8	15]	1, 3, 6, 9, 9, 8	16]	1, 3, 6, 9, 10, 8
17]	1, 3, 6, 9, 11, 8	18]	1, 3, 6, 9, 12, 8	21]	1, 3, 6, 10, 9, 8
22]	1, 3, 6, 10, 10, 8	23]	1, 3, 6, 10, 11, 8	24]	1, 3, 6, 10, 12, 8
25]	1, 3, 6, 10, 13, 8	26]	1, 3, 6, 10, 14, 8	27]	1, 3, 6, 10, 15, 8

1]. The lex segment ideal with this h-vector defines a level algebra. This is an example of a "minimal" level algebra (see [**5**]).

Notice that by **Corollary 5.17** 1], 3], 7], and 10] are the h-vectors of Artinian level algebras with the WLP.

We can apply **Proposition 5.24** to get level Artinian algebras with the WLP from **Table 5.7** for the following: 11] from 7], 12] from 8], 15] from 18], 16] from 19], 17] from 20], 21] from 23], 22] from 24], 23] from 25], 24] from 26], 25] from 27], 26] from 35], 27] from 36].

4] arises as the h-vector of 33 general points on a smooth ACM curve of degree 9 whose Hilbert function has first difference $(1,3,5,7,9,9,\ldots)$. Using **Proposition 6.17** (see **Example 6.18**) its Artinian reduction has the WLP.

The last remaining case, 18] arose as the h-vector of 39 points on a smooth cubic surface in $\mathbb{P}^3$. A computer check revealed that the Artinian reduction of the homogeneous coordinate ring of this example also satisfies the WLP.

Socle Degree 5 and Type 9

There are 20 examples in this case. They are in the table below.

Table 5.9A

1]	1, 3, 5, 7, 8, 9	2]	1, 3, 5, 7, 9, 9	3]	1, 3, 6, 7, 8, 9
4]	1, 3, 6, 7, 9, 9	5]	1, 3, 6, 8, 8, 9	6]	1, 3, 6, 8, 9, 9
7]	1, 3, 6, 8, 10, 9	8]	1, 3, 6, 9, 8, 9	9]	1, 3, 6, 9, 9, 9
10]	1, 3, 6, 9, 10, 9	11]	1, 3, 6, 9, 11, 9	12]	1, 3, 6, 9, 12, 9
13]	1, 3, 6, 10, 8, 9	14]	1, 3, 6, 10, 9, 9	15]	1, 3, 6, 10, 10, 9
16]	1, 3, 6, 10, 11, 9	17]	1, 3, 6, 10, 12, 9	18]	1, 3, 6, 10, 13, 9
19]	1, 3, 6, 10, 14, 9	20]	1, 3, 6, 10, 15, 9		

Non-existence

Eliminated using Corollary 3.5. 4].

Not cancelable, but not using any of our theorems. 5], 8], 13].

Eliminated using an argument completely analogous to Example 3.11. 14].

Existence

This leaves the following possible h-vectors.

Table 5.9

1]	1, 3, 5, 7, 8, 9	2]	1, 3, 5, 7, 9, 9	3]	1, 3, 6, 7, 8, 9
6]	1, 3, 6, 8, 9, 9	7]	1, 3, 6, 8, 10, 9	9]	1, 3, 6, 9, 9, 9
10]	1, 3, 6, 9, 10, 9	11]	1, 3, 6, 9, 11, 9	12]	1, 3, 6, 9, 12, 9
15]	1, 3, 6, 10, 10, 9	16]	1, 3, 6, 10, 11, 9	17]	1, 3, 6, 10, 12, 9
18]	1, 3, 6, 10, 13, 9	19]	1, 3, 6, 10, 14, 9	20]	1, 3, 6, 10, 15, 9

1]. This is the h-vector of a minimal level algebra, so the lex-segment ideal with this h-vector already describes a level algebra.

Notice that, by **Corollary 5.17**, 1], 2], 3], 6] and 9] are the h-vectors of Artinian level algebras with the WLP. By **Proposition 5.16**, 12] is also the h-vector of an Artinian level algebra with the WLP.

We can apply **Proposition 5.24** to get level Artinian algebras with the WLP from **Table 5.8** for the following: 7] from 4], 10] from 11], 11] from 12], 15] from 21], 16] from 22], 17] from 23], 18] from 24], 19] from 25], 20] from 26].

Socle Degree 5 and Type 10

There are 15 possible h-vectors. They are:

Table 5.10A

1]	1, 3, 5, 7, 9, 10	2]	1, 3, 6, 7, 9, 10	3]	1, 3, 6, 8, 9, 10
4]	1, 3, 6, 8, 10, 10	5]	1, 3, 6, 9, 9, 10	6]	1, 3, 6, 9, 10, 10
7]	1, 3, 6, 9, 11, 10	8]	1, 3, 6, 9, 12, 10	9]	1, 3, 6, 10, 9, 10
10]	1, 3, 6, 10, 10, 10	11]	1, 3, 6, 10, 11, 10	12]	1, 3, 6, 10, 12, 10
13]	1, 3, 6, 10, 13, 10	14]	1, 3, 6, 10, 14, 10	15]	1, 3, 6, 10, 15, 10

Non-existence

Eliminated using Corollary 3.5. 2].
Eliminated because not cancelable, but not using any of our theorems. 5].
We can eliminate 9] by using **Proposition 5.24** and the fact that 10] of **Table 5.11A** does not exist.

Existence

This leaves the following as the possible h-vectors of level artinian algebras.

Table 5.10

1]	1, 3, 5, 7, 9, 10	3]	1, 3, 6, 8, 9, 10	4]	1, 3, 6, 8, 10, 10
6]	1, 3, 6, 9, 10, 10	7]	1, 3, 6, 9, 11, 10	8]	1, 3, 6, 9, 12, 10
10]	1, 3, 6, 10, 10, 10	11]	1, 3, 6, 10, 11, 10	12]	1, 3, 6, 10, 12, 10
13]	1, 3, 6, 10, 13, 10	14]	1, 3, 6, 10, 14, 10	15]	1, 3, 6, 10, 15, 10

1]. This is the h-vector of a minimal level algebra, so the lex-segment ideal with this h-vector already describes a level algebra.

Notice that **Corollary 5.17** shows that 1], 3], 4], 6] and 10] are the h-vectors of Artinian level algebras with the WLP, while **Proposition 5.16** gives the same result for 11], 12], 13] and 14].

We can apply **Proposition 5.24** to get level Artinian algebras with the WLP from **Table 5.9** for the following: 7] from 7], 15] from 19].

The last remaining case, 8], arises as the h-vector of 41 points on a smooth cubic surface in $\mathbb{P}^3$. A computer check showed the Artinian reduction of the homogeneous coordinate ring of these points also satisfies the WLP.

Socle Degree 5 and Type 11

In this case there are 15 possible h-vectors to consider.

Table 5.11A

1]	1, 3, 5, 7, 9, 11	2]	1, 3, 6, 7, 9, 11	3]	1, 3, 6, 8, 9, 11
4]	1, 3, 6, 8, 10, 11	5]	1, 3, 6, 9, 9, 11	6]	1, 3, 6, 9, 10, 11
7]	1, 3, 6, 9, 11, 11	8]	1, 3, 6, 9, 12, 11	9]	1, 3, 6, 10, 9, 11
10]	1, 3, 6, 10, 10, 11	11]	1, 3, 6, 10, 11, 11	12]	1, 3, 6, 10, 12, 11
13]	1, 3, 6, 10, 13, 11	14]	1, 3, 6, 10, 14, 11	15]	1, 3, 6, 10, 15, 11

Non-existence

Eliminated using Corollary 3.5. 2], 3], 5], 9].
Eliminated by an argument completely analogous to that of Example 3.11. 10].

Existence

This leaves the following to consider.

Table 5.11

1]	1, 3, 5, 7, 9, 11	4]	1, 3, 6, 8, 10, 11	6]	1, 3, 6, 9, 10, 11
7]	1, 3, 6, 9, 11, 11	8]	1, 3, 6, 9, 12, 11	11]	1, 3, 6, 10, 11, 11
12]	1, 3, 6, 10, 12, 11	13]	1, 3, 6, 10, 13, 11	14]	1, 3, 6, 10, 14, 11
15]	1, 3, 6, 10, 15, 11				

1] This is the h-vector of a minimal level algebra, so the lex-segment ideal with this h-vector already describes a level algebra.

Notice that **Corollary 5.17** shows that 1], 4], 6], 7] and 11] are the h-vectors of Artinian level algebras with the WLP.

We can apply **Proposition 5.24** to get level Artinian algebras with the WLP from **Table 5.10** for the following: 12] from 11], 13] from 12], 14] from 13], 15] from 14].

The last remaining case, 8], corresponds to 42 points on a smooth cubic surface in $\mathbb{P}^3$. We showed, using a computer, that the Artinian reduction of the homogeneous coordinate ring of these points also satisfies the WLP.

Socle Degree 5 and Type 12

There are 10 possible h-vectors to consider.

Table 5.12A

1]	1, 3, 6, 8, 10, 12	2]	1, 3, 6, 9, 10, 12	3]	1, 3, 6, 9, 11, 12
4]	1, 3, 6, 9, 12, 12	5]	1, 3, 6, 10, 10, 12	6]	1, 3, 6, 10, 11, 12
7]	1, 3, 6, 10, 12, 12	8]	1, 3, 6, 10, 13, 12	9]	1, 3, 6, 10, 14, 12
10]	1, 3, 6, 10, 15, 12				

Non-existence

Eliminated by Corollary 3.5. 2], 5].

Existence

This leaves the following as the possible h-vector of a level algebra.

Table 5.12

1]	1, 3, 6, 8, 10, 12	3]	1, 3, 6, 9, 11, 12	4]	1, 3, 6, 9, 12, 12
6]	1, 3, 6, 10, 11, 12	7]	1, 3, 6, 10, 12, 12	8]	1, 3, 6, 10, 13, 12
9]	1, 3, 6, 10, 14, 12	10]	1, 3, 6, 10, 15, 12		

1]. This is the h-vector of a minimal level algebra, so the lex-segment ideal with this h-vector already describes a level algebra.

Notice that **Corollary 5.17** gives that 1], 3], 4], 6] and 7] are the h-vectors of Artinian level algebras with the WLP.

We can apply **Proposition 5.24** to get level Artinian algebras with the WLP from **Table 5.11** for the three remaining examples as follows: 8] from 12], 9] from 13], 10] from 14].

Socle Degree 5 and Type 13

In this case we have 7 vectors to consider.

Table 5.13A

1]	1, 3, 6, 9, 11, 13	2]	1, 3, 6, 9, 12, 13	3]	1, 3, 6, 10, 11, 13
4]	1, 3, 6, 10, 12, 13	5]	1, 3, 6, 10, 13, 13	6]	1, 3, 6, 10, 14, 13
7]	1, 3, 6, 10, 15, 13				

Non-existence

Eliminated because not cancelable, but not by any of our theorems. 3].

Existence

This leaves the following as the possible h-vectors of a level algebra.

Table 5.13

1]	1, 3, 6, 9, 11, 13	2]	1, 3, 6, 9, 12, 13	4]	1, 3, 6, 10, 12, 13
5]	1, 3, 6, 10, 13, 13	6]	1, 3, 6, 10, 14, 13	7]	1, 3, 6, 10, 15, 13

1]. This is the h-vector of a minimal level algebra, so the lex-segment ideal with this h-vector already describes a level algebra.

Notice that by **Corollary 5.17**, 1], 2], 4] and 5] are the h-vectors of Artinian level algebras with the WLP.

We can apply **Proposition 5.24** to get level Artinian algebras with the WLP from **Table 5.12** for the remaining two cases as follows: 6] from 8], 7] from 9].

Socle Degree 5 and Type 14

There are 5 vectors to consider. None can be eliminated

Table 5.14

1]	1, 3, 6, 9, 12, 14	2]	1, 3, 6, 10, 12, 14	3]	1, 3, 6, 10, 13, 14
4]	1, 3, 6, 10, 14, 14	5]	1, 3, 6, 10, 15, 14		

1]. This is the h-vector of a minimal level algebra, so the lex-segment ideal with this h-vector already describes a level algebra.

Notice that by **Corollary 5.17**, 1], 2], 3] and 4] are the h-vectors of Artinian level algebras with the WLP.

The last remaining case 5], arises as the h-vector of an Artinian level algebra with the WLP by using **Proposition 5.24** on 6] in **Table 5.13**.

Socle Degree 5 and Type 15

There are 5 vectors to consider.

Table 5.15A

1]	1, 3, 6, 9, 12, 15	2]	1, 3, 6, 10, 12, 15	3]	1, 3, 6, 10, 13, 15
4]	1, 3, 6, 10, 14, 15	5]	1, 3, 6, 10, 15, 15		

Non-existence

Eliminated by Corollary 3.5. 2].

Existence

This leaves the following as the possible h-vector of a level algebra.

Table 5.15

1]	1, 3, 6, 9, 12, 15	3]	1, 3, 6, 10, 13, 15	4]	1, 3, 6, 10, 14, 15
5]	1, 3, 6, 10, 15, 15				

1]. This is the h-vector of a minimal level algebra, so the lex-segment ideal with this h-vector already describes a level algebra. By **Corollary 5.17**, all can be constructed as the h-vectors of Artinian level algebras with the WLP.

Socle Degree 5 and Type 16

There are only 3 possible vectors to consider. None can be eliminated.

Table 5.16

1]	1, 3, 6, 10, 13, 16	2]	1, 3, 6, 10, 14, 16	3]	1, 3, 6, 10, 15, 16

By **Corollary 5.17**, each is the h-vector of an Artinian level algebra with the WLP.

Socle Degree 5 and Type 17

There are only two possible h-vectors of level algebras. They are:

Table 5.17

1]	1, 3, 6, 10, 14, 17	2]	1, 3, 6, 10, 15, 17

By **Corollary 5.17** both are the h-vector of an Artinian level algebra with the WLP.

Socle Degree 5 and Type 18

There are only two possible h-vectors of level algebras. They are:

Table 5.18

1]	1, 3, 6, 10, 14, 18	2]	1, 3, 6, 10, 15, 18

By **Corollary 5.17**, both are the h-vectors of an Artinian level algebra with the WLP.

Socle Degree 5 and Type 19

There is only one h-vector possible here. It is the h-vector of both the compressed and minimal level algebra of its type. It is:

Table 5.19

1]	1, 3, 6, 10, 15, 19

and by **Corollary 5.17** is the h-vector of an Artinian level algebra with the WLP.

Socle Degree 5 and Type 20

There is only one h-vector possible here. It is the h-vector of both the compressed and minimal level algebra of its type. It is:

Table 5.20

1]	1, 3, 6, 10, 15, 20

and by **Corollary 5.17** is the h-vector of an Artinian level algebra with the WLP.

Socle Degree 5 and Type 21

There is only one h-vector possible here. It is the h-vector of both the compressed and minimal level algebra of its type. It is:

Table 5.21

1]	1, 3, 6, 10, 15, 21

and by **Corollary 5.17** is the h-vector of an Artinian level algebra with the WLP.

APPENDIX D

Socle Degree 4

Socle Degree 4 and Type 2

This case was treated in our paper [**23**]. We showed that the following are the h-vectors of level Artinian algebras of socle degree 4 and type 2.

Table 4.2

1]	1, 3, 3, 3, 2	2]	1, 3, 4, 3, 2	3]	1, 3, 4, 4, 2	4]	1, 3, 4, 5, 2
5]	1, 3, 5, 4, 2	6]	1, 3, 5, 5, 2	7]	1, 3, 5, 6, 2	8]	1, 3, 6, 4, 2
9]	1, 3, 6, 5, 2	10]	1, 3, 6, 6, 2				

Of the examples found in [[**23**], Example 3.18] one easily sees that 2], 3], 5], 7] and 8] also have the WLP by applying **Proposition 5.15**. The remaining 5 h-vectors can also be shown to be the h-vectors of level Artinian algebras with the WLP by applying **Proposition 5.18** to the appropriate examples from **Table 5.2**.

Points

We show that each of the h-vectors in Table 4.2 is also the h-vector of a set of level points in $\mathbb{P}^3$.

1] This is the h-vector of 12 points on a twisted cubic curve in $\mathbb{P}^3$. That is a level set of points.

2] Start with 4 collinear points and perform general links of type (respectively) $(3, 3, 4)$ and $(3, 3, 5)$.

3] Start with two points and perform a general link of type $(2, 2, 4)$.

4] Start with two points and perform general links of type (respectively) $(1, 3, 3)$, $(2, 3, 4)$, $(2, 4, 4)$.

5] The diagram in [[**23**], Example 3.18] lifts, since the points have the "nested" property.

6] Start with two points in $\mathbb{P}3$ and link in a general complete intersection of type $(2, 3, 3)$.

7] Start with 3 general points and link in general complete intersections of type (respectively) $(2, 3, 3)$, $(2, 4, 4)$.

8] Start with 3 collinear points and link in complete intersections of type (respectively) $(2, 2, 3)$, $(2, 2, 4)$, $(3, 3, 4)$, $(3, 3, 5)$.

9] Start with a complete intersection of type $(1, 2, 2)$ and link with general complete intersections of type (respectively) $(2, 2, 3)$, $(3, 3, 3)$, $(3, 3, 4)$.

10] Take 18 general points on a general arithmetically Cohen-Macaulay curve of degree 6 and genus 3.

Socle Degree 4 and Type 3

There are 17 possible Artinian h-vectors, listed in **Table 4.3A**.

Table 4.3A

1]	1, 3, 3, 3, 3	2]	1, 3, 3, 4, 3	3]	1, 3, 4, 3, 3	4]	1, 3, 4, 4, 3
5]	1, 3, 4, 5, 3	6]	1, 3, 5, 3, 3	7]	1, 3, 5, 4, 3	8]	1, 3, 5, 5, 3
9]	1, 3, 5, 6, 3	10]	1, 3, 5, 7, 3	11]	1, 3, 6, 3, 3	12]	1, 3, 6, 4, 3
13]	1, 3, 6, 5, 3	14]	1, 3, 6, 6, 3	15]	1, 3, 6, 7, 3	16]	1, 3, 6, 8, 3
17]	1, 3, 6, 9, 3						

Non-existence

Eliminated by Corollary 3.5. 2].
Eliminated by Proposition 3.8, b). $d = 3$, 3], 6], 11].

Eliminated by an argument completely analogous to Example 3.11. 12].

Existence

This leaves us with the h-vectors in **Table 4.3**, which we show are all the h-vectors of Artinian level algebras with the WLP and the h-vectors of level sets of points in $\mathbb{P}^3$.

Table 4.3

1]	1, 3, 3, 3, 3	4]	1, 3, 4, 4, 3	5]	1, 3, 4, 5, 3	7]	1, 3, 5, 4, 3
8]	1, 3, 5, 5, 3	9]	1, 3, 5, 6, 3	10]	1, 3, 5, 7, 3	13]	1, 3, 6, 5, 3
14]	1, 3, 6, 6, 3	15]	1, 3, 6, 7, 3	16]	1, 3, 6, 8, 3	17]	1, 3, 6, 9, 3

Note that 1] and 4] come from socle degree 5 and type 2 (by truncation) and so are already known to be the h-vectors of level sets of points in $\mathbb{P}^3$ and also of Artinian level algebras with the WLP. The same is true for 8] and 14] by **Proposition 5.16**.

The Link-Sum Construction All the remaining examples (except 17]) can be made by the linked-sum method. All of these cases (**except 7] and 13]**) satisfy the WLP by **Proposition 5.15**. But, 7], 13] and 17] can be shown to be the h-vectors of level Artinian algebras with the WLP by using **Proposition 5.18** on the appropriate examples in **Table 5.3**. We use the (by now) usual linked-sum notation.

5]. $\mathbb{Z} = \begin{Bmatrix} * & & \\ * & * & \\ * & \bullet & \bullet \\ * & \bullet & \bullet \\ \bullet & \bullet & \bullet \\ \bullet & \bullet & \bullet \end{Bmatrix}$
$\begin{array}{llllllll} \mathbf{H}(\mathbb{Z}, -) & : & 1 & 3 & 6 & 9 & 12 & 15 & \to \\ \mathbf{H}(\mathbb{X}, -) & : & 1 & 3 & 6 & 9 & 10 & 10 & \to \\ \mathbf{H}(\mathbb{Y}, -) & : & 1 & 3 & 4 & 5 & 5 & 5 & \to \\ \mathbf{H}(A, -) & : & 1 & 3 & 4 & 5 & 3 & 0 & \to. \end{array}$

D. SOCLE DEGREE 4

7]. $\mathbb{Z} = \left\{\begin{matrix} \bullet & & \\ \bullet & \bullet & \\ \bullet & \bullet & \bullet \\ \bullet & \bullet & * \\ \bullet & \bullet & * \\ * & * & * \end{matrix}\right\}$
$\mathbf{H}(\mathbb{Z},-)$: 1 3 6 9 12 15 $\to$
$\mathbf{H}(\mathbb{X},-)$: 1 3 6 8 10 10 $\to$
$\mathbf{H}(\mathbb{Y},-)$: 1 3 5 5 5 5 $\to$
$\mathbf{H}(A,-)$: 1 3 5 4 3 0 $\to$.

9]. $\mathbb{Z} = \left\{\begin{matrix} \bullet & & \\ \bullet & \bullet & \\ \bullet & \bullet & \bullet \\ \bullet & \bullet & \bullet \\ * & * & * \\ * & * & * \end{matrix}\right\}$
$\mathbf{H}(\mathbb{Z},-)$: 1 3 6 9 12 15 $\to$
$\mathbf{H}(\mathbb{X},-)$: 1 3 6 9 9 9 $\to$
$\mathbf{H}(\mathbb{Y},-)$: 1 3 5 6 6 6 $\to$
$\mathbf{H}(A,-)$: 1 3 5 6 3 0 $\to$.

10]. $\mathbb{Z} = \left\{\begin{matrix} \bullet & & & \\ * & & & \\ * & \bullet & \bullet & \\ * & \bullet & \bullet & \bullet \\ \bullet & * & \bullet & \bullet \\ * & * & \bullet & \bullet \\ * & * & \bullet & \bullet \end{matrix}\right\}$
$\mathbf{H}(\mathbb{Z},-)$: 1 3 6 10 14 17 $\to$
$\mathbf{H}(\mathbb{X},-)$: 1 3 6 10 10 10 $\to$
$\mathbf{H}(\mathbb{Y},-)$: 1 3 5 7 7 7 $\to$
$\mathbf{H}(A,-)$: 1 3 5 7 3 0 $\to$.

13]. $\mathbb{Z} = \left\{\begin{matrix} \bullet & & \\ \bullet & \bullet & \\ * & \bullet & * \\ \bullet & * & * \\ \bullet & \bullet & \bullet \\ \bullet & * & * \end{matrix}\right\}$
$\mathbf{H}(\mathbb{Z},-)$: 1 3 6 9 12 15 $\to$
$\mathbf{H}(\mathbb{X},-)$: 1 3 6 8 9 9 $\to$
$\mathbf{H}(\mathbb{Y},-)$: 1 3 6 6 6 6 $\to$
$\mathbf{H}(A,-)$: 1 3 6 5 3 0 $\to$.

15]. $\mathbb{Z} = \left\{\begin{matrix} \bullet & \bullet & \bullet & & \\ \bullet & * & * & * & \\ \bullet & * & * & \bullet & \bullet \\ \bullet & \bullet & * & * & \bullet \end{matrix}\right\}$
$\mathbf{H}(\mathbb{Z},-)$: 1 3 6 10 14 17 $\to$
$\mathbf{H}(\mathbb{X},-)$: 1 3 6 10 10 10 $\to$
$\mathbf{H}(\mathbb{Y},-)$: 1 3 6 7 7 7 $\to$
$\mathbf{H}(A,-)$: 1 3 6 7 3 0 $\to$.

16]. $\mathbb{Z} = \left\{\begin{matrix} \bullet & \bullet & & & \\ * & \bullet & & & \\ \bullet & * & \bullet & * & \\ * & \bullet & * & \bullet & \bullet \\ \bullet & \bullet & * & * & * \end{matrix}\right\}$
$\mathbf{H}(\mathbb{Z},-)$: 1 3 6 10 15 18 $\to$
$\mathbf{H}(\mathbb{X},-)$: 1 3 6 10 10 10 $\to$
$\mathbf{H}(\mathbb{Y},-)$: 1 3 6 8 8 8 $\to$
$\mathbf{H}(A,-)$: 1 3 6 8 3 0 $\to$.

17]. This is the h-vector of a compressed level algebra.

Points

5]. As points in $\mathbb{P}^3$, this can be obtained by lifting the above diagram.
7]. As points in $\mathbb{P}^3$, this can be obtained by lifting the above diagram.
9]. As points in $\mathbb{P}^3$, this can be obtained by lifting the above diagram.
10]. As points in $\mathbb{P}^3$, this can be obtained as a set of 19 general points on a smooth curve of type $(3, 4)$ on a smooth quadric surface.

13]. As points in $\mathbb{P}^3$, this can be obtained by starting with three collinear points and linking with general complete intersections of type (respectively) $(2, 2, 3)$, $(2, 3, 4)$, $(3, 3, 4)$, $(3, 4, 4)$, $(3, 3, 5)$.

For the next three cases we follow the following construction. Consider a general arithmetically Gorenstein set of points, $\mathbb{X}$, with h-vector $(1, 3, 6, a, 6, 3, 1)$ where $a = 7$, 8 or 9. (The case $a = 7$ is a complete intersection of type $(3, 3, 3)$.) One can find inside $\mathbb{X}$ a set of points, $\mathbb{Z}$, with h-vector $(1, 3, 3)$ which, furthermore, has three generators in degree 2 and precisely one in degree 3. When we link $\mathbb{Z}$ using the arithmetically Gorenstein set of points $\mathbb{X}$, we split off a summand corresponding to the cubic generator of $\mathbb{Z}$, and the residual set of points is level with the desired h-vector for cases 15], 16], and 17].

Socle Degree 4 and Type 4

There are 14 possible h-vectors, listed in **Table 4.4A**.

Table 4.4A

1]	1, 3, 3, 4, 4	2]	1, 3, 4, 4, 4	3]	1, 3, 4, 5, 4	4]	1, 3, 5, 4, 4
5]	1, 3, 5, 5, 4	6]	1, 3, 5, 6, 4	7]	1, 3, 5, 7, 4	8]	1, 3, 6, 4, 4
9]	1, 3, 6, 5, 4	10]	1, 3, 6, 6, 4	11]	1, 3, 6, 7, 4	12]	1, 3, 6, 8, 4
13]	1, 3, 6, 9, 4	14]	1, 3, 6, 10, 4					

Non-existence

Eliminated by Corollary 3.5. 1].

Eliminated by Proposition 3.8, c). $d + 1 = 4$, 4], 8].

Existence

This leaves the h-vectors in **Table 4.4**. We show that all of them are the h-vectors of level Artinian algebras with the WLP and also the h-vectors of level sets of points in $\mathbb{P}^3$.

Table 4.4

2]	1, 3, 4, 4, 4	3]	1, 3, 4, 5, 4	5]	1, 3, 5, 5, 4	6]	1, 3, 5, 6, 4
7]	1, 3, 5, 7, 4	9]	1, 3, 6, 5, 4	10]	1, 3, 6, 6, 4	11]	1, 3, 6, 7, 4
12]	1, 3, 6, 8, 4	13]	1, 3, 6, 9, 4	14]	1, 3, 6, 10, 4		

Note that 2], 3], 5], 6], 10], and 11] all come from socle degree 5 and type 2 (by truncation) and so are automatically the h-vectors of level Artinian algebras with the WLP and also of level sets of points in $\mathbb{P}^3$.

The Link-Sum Construction We make all the remaining cases (except 14]) using the linked sum method (with the standard notation). By **Proposition 5.15** we can see that all the examples satisfy the WLP. To see that 14] is also the h-vector of a level Artinian algebra with the WLP we apply **Proposition 5.18** to the appropriate example in **Table 5.4**.

D. SOCLE DEGREE 4   123

7]. $\mathbb{Z} = \left\{\begin{matrix} * & & \\ \bullet & \bullet & \\ \bullet & \bullet & * \\ * & \bullet & * & \bullet \\ * & \bullet & \bullet & \bullet \\ * & \bullet & * & \bullet \end{matrix}\right\}$
$\mathbf{H}(\mathbb{Z},-)$: 1 3 6 10 14 18 →
$\mathbf{H}(\mathbb{X},-)$: 1 3 6 10 11 11 →
$\mathbf{H}(\mathbb{Y},-)$: 1 3 5 7   7   7   →
$\mathbf{H}(A,-)$: 1 3 5 7   4   0   → .

9]. $\mathbb{Z} = \left\{\begin{matrix} * & & & \\ \bullet & \bullet & & \\ \bullet & \bullet & * & \\ \bullet & \bullet & \bullet & * \\ \bullet & \bullet & \bullet & * \\ \bullet & * & \bullet & * \end{matrix}\right\}$
$\mathbf{H}(\mathbb{Z},-)$: 1 3 6 10 14 18 →
$\mathbf{H}(\mathbb{X},-)$: 1 3 6 9   12 12 →
$\mathbf{H}(\mathbb{Y},-)$: 1 3 6 6   6   6   →
$\mathbf{H}(A,-)$: 1 3 6 5   4   0   → .

12]. $\mathbb{Z} = \left\{\begin{matrix} \bullet & & & \\ * & * & & \\ * & \bullet & * & \\ \bullet & * & \bullet & \bullet \\ * & \bullet & \bullet & \bullet \\ * & \bullet & * & \bullet \end{matrix}\right\}$
$\mathbf{H}(\mathbb{Z},-)$: 1 3 6 10 14 18 →
$\mathbf{H}(\mathbb{X},-)$: 1 3 6 10 10 10 →
$\mathbf{H}(\mathbb{Y},-)$: 1 3 6 8   8   8   →
$\mathbf{H}(A,-)$: 1 3 6 8   4   0   → .

13]. $\mathbb{Z} = \left\{\begin{matrix} \bullet & * & & & \\ \bullet & \bullet & * & & \\ \bullet & \bullet & \bullet & & \\ * & * & * & \bullet & * \\ * & \bullet & * & \bullet & * & \bullet \end{matrix}\right\}$
$\mathbf{H}(\mathbb{Z},-)$: 1 3 6 10 15 19 →
$\mathbf{H}(\mathbb{X},-)$: 1 3 6 10 10 10 →
$\mathbf{H}(\mathbb{Y},-)$: 1 3 6 9   9   9   →
$\mathbf{H}(A,-)$: 1 3 6 9   4   0   → .

14]. This is the h-vector of a compressed level algebra.

Points

7]. As points in $\mathbb{P}^3$ this can be obtained by starting with the union, $\mathbb{Z}$, of three collinear points and three general points, and then linking using complete intersections of type (respectively) $(2,3,3)$, $(2,4,4)$.

9]. As points inn $\mathbb{P}^3$, this can be obtained by starting with the union, $\mathbb{Z}$, of 4 collinear points and two more general points, and then linking using complete intersections of type (respectively) $(2,2,4)$, $(3,3,4)$, $(3,3,5)$.

12]. As points in $\mathbb{P}^3$, this can be obtained by taking 22 general points on a general arithmetically Cohen-Macaulay curve of degree 8 and genus 7.

13]. As points in $\mathbb{P}^3$, this can be obtained by taking 23 general points on a general arithmetically Cohen-Macaulay curve of degree 9 and genus 9.

14]. As points in $\mathbb{P}^3$, this can be obtained by taking 24 general points in $\mathbb{P}^3$.

Socle Degree 4 and Type 5

The fourteen possible h-vectors in this case, are listed in **Table 4.5A**.

D. SOCLE DEGREE 4

Table 4.5A

1]	1, 3, 3, 4, 5	2]	1, 3, 4, 4, 5	3]	1, 3, 4, 5, 5	4]	1, 3, 5, 4, 5
5]	1, 3, 5, 5, 5	6]	1, 3, 5, 6, 5	7]	1, 3, 5, 7, 5	8]	1, 3, 6, 4, 5
9]	1, 3, 6, 5, 5	10]	1, 3, 6, 6, 5	11]	1, 3, 6, 7, 5	12]	1, 3, 6, 8, 5
13]	1, 3, 6, 9, 5	14]	1, 3, 6, 10, 5				

Non-existence

Eliminated by Corollary 3.5. 1], 2], 4], 8].
Eliminated by an argument completely analogous to Example 3.11. 9].

Existence

All the remaining h-vectors (listed in **Table 4.5**) are the h-vectors of Artinian level algebras with the WLP and also of level sets of points in $\mathbb{P}^3$.

Table 4.5

3]	1, 3, 4, 5, 5	5]	1, 3, 5, 5, 5	6]	1, 3, 5, 6, 5	7]	1, 3, 5, 7, 5
10]	1, 3, 6, 6, 5	11]	1, 3, 6, 7, 5	12]	1, 3, 6, 8, 5	13]	1, 3, 6, 9, 5
14]	1, 3, 6, 10, 5						

Of the h-vectors listed in **Table 4.5**, we can obtain almost all from socle degree 5 and type 2 (by truncation). The h-vectors so obtained are: 3], 5], 6], 7], 10], 11], 12], 13]. These, then, are automatically the h-vectors of Artinian level algebras with the WLP and also of level sets of points in $\mathbb{P}^3$.

This leaves only:

14]. This is the h-vector of a compressed level algebra and the h-vector of 25 general points in $\mathbb{P}^3$. The fact that this is also the h-vector of a level Artinian algebra with the WLP comes by applying **Proposition 5.18** to the appropriate example in **Table 5.5**.

Socle Degree 4 and Type 6

There are 10 possible h-vectors in this case and they are listed in **Table 4.6A**.

Table 4.6A

1]	1, 3, 4, 5, 6	2]	1, 3, 5, 5, 6	3]	1, 3, 5, 6, 6	4]	1, 3, 5, 7, 6
5]	1, 3, 6, 5, 6	6]	1, 3, 6, 6, 6	7]	1, 3, 6, 7, 6	8]	1, 3, 6, 8, 6
9]	1, 3, 6, 9, 6	10]	1, 3, 6, 10, 6				

Non-existence

Eliminated by Corollary 3.5. 2], 5].

Existence

This leaves the h-vectors in **Table 4.6**.

Table 4.6

1]	1, 3, 4, 5, 6	3]	1, 3, 5, 6, 6	4]	1, 3, 5, 7, 6	6]	1, 3, 6, 6, 6
7]	1, 3, 6, 7, 6	8]	1, 3, 6, 8, 6	9]	1, 3, 6, 9, 6	10]	1, 3, 6, 10, 6

Socle Degree 4 and Type 7

In this case we only have 7 possible h-vectors, given in **Table 4.7A**.

Table 4.7A

1]	1, 3, 5, 6, 7	2]	1, 3, 5, 7, 7	3]	1, 3, 6, 6, 7	4]	1, 3, 6, 7, 7
5]	1, 3, 6, 8, 7	6]	1, 3, 6, 9, 7	7]	1, 3, 6, 10, 7		

Non-existence

Not cancelable, but not using any of our theorems. 3].

Existence

All the remaining h-vectors are the h-vector of Artinian level algebras with the WLP and h-vectors of level sets of points in $\mathbb{P}^3$, and are listed in **Table 4.7**.

Table 4.7

1]	1, 3, 5, 6, 7	2]	1, 3, 5, 7, 7	4]	1, 3, 6, 7, 7
5]	1, 3, 6, 8, 7	6]	1, 3, 6, 9, 7	7]	1, 3, 6, 10, 7

All of these examples come from socle degree 6 and type 2 (by truncation).

Socle Degree 4 and Type 8

All the 5 possible h-vectors are listed in **Table 4.8** and all of them are the h-vectors of level Artinian algebras with the WLP and the h-vectors of level sets of points in $\mathbb{P}^3$. All of these come from socle degree 6 and type 2 (by truncation).

Table 4.8

1]	1, 3, 5, 7, 8	2]	1, 3, 6, 7, 8	3]	1, 3, 6, 8, 8	4]	1, 3, 6, 9, 8
5]	1, 3, 6, 10, 8						

Socle Degree 4 and Type 9

The 5 possible h-vectors are listed in **Table 4.9A**.

Table 4.9A

1]	1, 3, 5, 7, 9	2]	1, 3, 6, 7, 9	3]	1, 3, 6, 8, 9	4]	1, 3, 6, 9, 9
5]	1, 3, 6, 10, 9						

Non-existence

Eliminated by Corollary 3.5. 2].

Existence

The remaining h-vectors, listed in **Table 4.9**, are all the h-vectors of level Artinian algebras with the WLP and the h-vectors of level sets of point in $\mathbb{P}^3$. All can be obtained from the examples in socle degree 6 and type 2 (by truncation).

Table 4.9

| 1] | 1, 3, 5, 7, 9 | 3] | 1, 3, 6, 8, 9 | 4] | 1, 3, 6, 9, 9 | 5] | 1, 3, 6, 10, 9 |

Socle Degree 4 and Type 10

There are only 3 possible h-vectors and all of these are the h-vectors of level Artinian algebras with the WLP and the h-vectors of level sets of point in $\mathbb{P}^3$. All can be obtained from the examples in socle degree 6 and type 2 (by truncation). These are listed in **Table 4.10**.

Table 4.10

| 1] | 1, 3, 6, 8, 10 | 2] | 1, 3, 6, 9, 10 | 3] | 1, 3, 6, 10, 10 |

Socle Degree 4 and Type 11

There are only 2 possible h-vectors and both of these are the h-vectors of level Artinian algebras with the WLP and the h-vectors of level sets of point in $\mathbb{P}^3$. All can be obtained from the examples in socle degree 6 and type 2 (by truncation). These are listed in **Table 4.11**.

Table 4.11

| 1] | 1, 3, 6, 9, 11 | 2] | 1, 3, 6, 10, 11 |

Socle Degree 4 and Type 12

There are only 2 possible h-vectors and both of these are the h-vectors of level Artinian algebras with the WLP and the h-vectors of level sets of point in $\mathbb{P}^3$. Both can be obtained from the examples in socle degree 6 and type 2 (by truncation). These are listed in **Table 4.12**.

Table 4.12

| 1] | 1, 3, 6, 9, 12 | 2] | 1, 3, 6, 10, 12 |

Socle Degree 4 and Type 13

There is only one possible h-vector in this case and it is:

Table 4.13

| 1] | 1, 3, 6, 10, 13 |

This is the h-vector of a minimal level algebra and hence the lex-segment ideal with this h-vector gives a level ring. The lift of that monomial ideal to points also gives us an example with points in $\mathbb{P}^3$. It is easy to find an example with the WLP.

Socle Degree 4 and Type 14

The discussion is exactly as in the previous case and we have only one example.

Table 4.14

| 1] | 1, 3, 6, 10, 14 |

Socle Degree 4 and Type 15

The discussion is exactly as in the previous case and we have only one example.

Table 4.15

| 1] | 1, 3, 6, 10, 15 |

APPENDIX E

Socle Degree 3

Socle Degree 3 and Type 2

In this case there are exactly 4 level h-vectors (this case was treated in [[**23**], Example 3.18]). They are:

Table 3.2

| 1] | 1, 3, 3, 2 | 2] | 1, 3, 4, 2 | 3] | 1, 3, 5, 2 | 4] | 1, 3, 6, 2 |

Artinian level algebras for each of these cases was constructed in [[**23**], Example 3.18]. It is easy to check that all have the WLP, using **Proposition 5.15**.

Points

1]. Take 9 points on a twisted cubic in $\mathbb{P}^3$.
2]. Take 10 general points on a complete intersection curve of $\mathbb{P}^3$ of type $(2,2)$.
3]. Start with 7 general points in $\mathbb{P}^3$ and link in a complete intersection of type $(2,3,3)$.
4]. Take 12 general points in $\mathbb{P}^3$.

Socle Degree 3 and Type 3

In this case there are exactly 4 possible Artinian h-vectors and they are each the h-vector of a level algebra.

Table 3.3

| 1] | 1, 3, 3, 3 | 2] | 1, 3, 4, 3 | 3] | 1, 3, 5, 3 | 4] | 1, 3, 6, 3 |

Notice that by **Proposition 5.16** all of these are the h-vectors of Artinian level algebras with the WLP and also the h-vectors of level sets of points in $\mathbb{P}^3$.

Socle Degree 3 and Type 4

There are 4 Artinian h-vectors. They are:

Table 3.4A

| 1] | 1, 3, 3, 4 | 2] | 1, 3, 4, 4 | 3] | 1, 3, 5, 4 | 4] | 1, 3, 6, 4 |

Non-existence

Eliminated by Corollary 3.5. 1].

Existence

The remaining are the h-vectors of level Artinian algebras with the WLP and also the h-vectors of points in $\mathbb{P}^3$. They can all be obtained from the examples in socle degree 4 and type 2 (by truncation).

Table 3.4

| 2] | 1, 3, 4, 4 | 3] | 1, 3, 5, 4 | 4] | 1, 3, 6, 4 |

Socle Degree 3 and Type 5

In this case there are only three possible Artinian h-vectors and all are the h-vectors of level algebras.

Table 3.5

| 1] | 1, 3, 4, 5 | 2] | 1, 3, 5, 5 | 3] | 1, 3, 6, 5 |

These can all be derived from the examples in socle degree 4 and type 2 (by truncation) and hence are the h-vectors of both level Artinian algebras with the WLP and of level sets of points in $\mathbb{P}^3$.

Socle Degree 3 and Type 6

In this case there are only two possible Artinian h-vectors.

Table 3.6

| 1] | 1, 3, 5, 6 | 2] | 1, 3, 6, 6 |

These can both be obtained from the examples in socle degree 6 and type 2 (by truncation). So, we have these both as the h-vectors of Artinian level algebras with the WLP and also as the h-vectors of level sets of points in $\mathbb{P}^3$.

Socle Degree 3 and Type 7

In this case there are only two possible Artinian h-vectors.

Table 3.7

| 1] | 1, 3, 5, 7 | 2] | 1, 3, 6, 7 |

These can both be obtained from the examples in socle degree 6 and type 2 (by truncation). So, we have these both as the h-vectors of Artinian level algebras with the WLP and also as the h-vectors of level sets of points in $\mathbb{P}^3$.

Socle Degree 3 and Type 8

In this case there is only one possible Artinian h-vector.

Table 3.8

| 1] | 1, 3, 6, 8 |

This can be obtained from the examples in socle degree 6 and type 2 (by truncation). So, we have this as the h-vector of an Artinian level algebra with the WLP and also as the h-vector of a level set of points in $\mathbb{P}^3$.

Socle Degree 3 and Type 9

In this case there is only one possible Artinian h-vector.

Table 3.9

| 1] | 1, 3, 6, 9 |

This can be obtained from the examples in socle degree 6 and type 2 (by truncation). So, we have this as the h-vector of an Artinian level algebra with the WLP and also as the h-vector of a level set of points in $\mathbb{P}^3$.

Socle Degree 3 and Type 10

In this case there is only one possible Artinian h-vector.

Table 3.10

| 1] | 1, 3, 6, 10 |

This can be obtained from the examples in socle degree 6 and type 2 (by truncation). So, we have this as the h-vector of an Artinian level algebra with the WLP and also as the h-vector of a level set of points in $\mathbb{P}^3$.

APPENDIX F

Summary

We now summarize the work we did in this appendix by listing the type vectors of all level h-vectors (of type > 1) in socle degree ≤ 5 and also of type 2 and socle degree 6.

Table 3.2

| 1, 3, 3, 2 | 1, 3, 4, 2 | 1, 3, 5, 2 | 1, 3, 6, 2 |

Table 3.3

| 1, 3, 3, 3 | 1, 3, 4, 3 | 1, 3, 5, 3 | 1, 3, 6, 3 |

Table 3.4

| 1, 3, 4, 4 | 1, 3, 5, 4 | 1, 3, 6, 4 |

Table 3.5

| 1, 3, 4, 5 | 1, 3, 5, 5 | 1, 3, 6, 5 |

Table 3.6

| 1, 3, 5, 6 | 1, 3, 6, 6 |

Table 3.7

| 1, 3, 5, 7 | 1, 3, 6, 7 |

Table 3.8

| 1, 3, 6, 8 |

Table 3.9

| 1, 3, 6, 9 |

Table 3.10

| 1, 3, 6, 10 |

Table 4.2

| 1, 3, 3, 3, 2 | 1, 3, 4, 3, 2 | 1, 3, 4, 4, 2 | 1, 3, 4, 5, 2 | 1, 3, 5, 4, 2 |
| 1, 3, 5, 5, 2 | 1, 3, 5, 6, 2 | 1, 3, 6, 4, 2 | 1, 3, 6, 5, 2 | 1, 3, 6, 6, 2 |

Table 4.3

| 1, 3, 3, 3, 3 | 1, 3, 4, 4, 3 | 1, 3, 4, 5, 3 | 1, 3, 5, 4, 3 | 1, 3, 5, 5, 3 | 1, 3, 5, 6, 3 |
| 1, 3, 5, 7, 3 | 1, 3, 6, 5, 3 | 1, 3, 6, 6, 3 | 1, 3, 6, 7, 3 | 1, 3, 6, 8, 3 | 1, 3, 6, 9, 3 |

Table 4.4

1, 3, 4, 4, 4	1, 3, 4, 5, 4	1, 3, 5, 5, 4	1, 3, 5, 6, 4	1, 3, 5, 7, 4
1, 3, 6, 5, 4	1, 3, 6, 6, 4	1, 3, 6, 7, 4	1, 3, 6, 8, 4	1, 3, 6, 9, 4
1, 3, 6, 10, 4				

Table 4.5

| 1, 3, 4, 5, 5 | 1, 3, 5, 5, 5 | 1, 3, 5, 6, 5 | 1, 3, 5, 7, 5 | 1, 3, 6, 6, 5 |
| 1, 3, 6, 7, 5 | 1, 3, 6, 8, 5 | 1, 3, 6, 9, 5 | 1, 3, 6, 10, 5 | |

Table 4.6

| 1, 3, 4, 5, 6 | 1, 3, 5, 6, 6 | 1, 3, 5, 7, 6 | 1, 3, 6, 6, 6 | 1, 3, 6, 7, 6 |
| 1, 3, 6, 8, 6 | 1, 3, 6, 9, 6 | 1, 3, 6, 10, 6 | | |

Table 4.7

1, 3, 5, 6, 7	1, 3, 5, 7, 7
1, 3, 6, 7, 7	1, 3, 6, 8, 7
1, 3, 6, 9, 7	1, 3, 6, 10, 7

Table 4.8

1, 3, 5, 7, 8	1, 3, 6, 7, 8
1, 3, 6, 8, 8	1, 3, 6, 9, 8
1, 3, 6, 10, 8	

Table 4.9

1, 3, 5, 7, 9	1, 3, 6, 8, 9
1, 3, 6, 9, 9	1, 3, 6, 10, 9

Table 4.10

1, 3, 6, 8, 10	1, 3, 6, 9, 10
1, 3, 6, 10, 10	

Table 4.11

1, 3, 6, 9, 11	1, 3, 6, 10, 11

Table 4.12

1, 3, 6, 9, 12	1, 3, 6, 10, 12

Table 4.13

1, 3, 6, 10, 13

Table 4.14

1, 3, 6, 10, 14

Table 4.15

1, 3, 6, 10, 15

Table 5.2

1, 3, 3, 3, 3, 2	1, 3, 4, 4, 3, 2	1, 3, 4, 4, 4, 2	1, 3, 4, 5, 4, 2	1, 3, 4, 5, 5, 2
1, 3, 5, 5, 4, 2	1, 3, 5, 5, 5, 2	1, 3, 5, 6, 4, 2	1, 3, 5, 6, 5, 2	1, 3, 5, 6, 6, 2
1, 3, 5, 7, 5, 2	1, 3, 5, 7, 6, 2	1, 3, 6, 6, 4, 2	1, 3, 6, 6, 5, 2	1, 3, 6, 6, 6, 2
1, 3, 6, 7, 4, 2	1, 3, 6, 7, 5, 2	1, 3, 6, 7, 6, 2	1, 3, 6, 8, 5, 2	1, 3, 6, 8, 6, 2
1, 3, 6, 9, 5, 2	1, 3, 6, 9, 6, 2	1, 3, 6, 10, 6, 2		

Table 5.3

1, 3, 3, 3, 3, 3	1, 3, 4, 4, 4, 3	1, 3, 4, 5, 4, 3	1, 3, 4, 5, 5, 3
1, 3, 4, 5, 6, 3	1, 3, 5, 5, 4, 3	1, 3, 5, 5, 5, 3	1, 3, 5, 6, 5, 3
1, 3, 5, 6, 6, 3	1, 3, 5, 6, 7, 3	1, 3, 5, 7, 5, 3	1, 3, 5, 7, 6, 3
1, 3, 5, 7, 7, 3	1, 3, 5, 7, 8, 3	1, 3, 5, 7, 9, 3	1, 3, 6, 6, 5, 3
1, 3, 6, 6, 6, 3	1, 3, 6, 7, 5, 3	1, 3, 6, 7, 6, 3	1, 3, 6, 7, 7, 3
1, 3, 6, 7, 8, 3	1, 3, 6, 8, 5, 3	1, 3, 6, 8, 6, 3	1, 3, 6, 8, 7, 3
1, 3, 6, 8, 8, 3	1, 3, 6, 8, 9, 3	1, 3, 6, 9, 6, 3	1, 3, 6, 9, 7, 3
1, 3, 6, 9, 8, 3	1, 3, 6, 9, 9, 3	1, 3, 6, 10, 6, 3	1, 3, 6, 10, 7, 3
1, 3, 6, 10, 8, 3	1, 3, 6, 10, 9, 3		

Table 5.4

1, 3, 4, 4, 4, 4	1, 3, 4, 5, 5, 4	1, 3, 4, 5, 6, 4	1, 3, 5, 5, 5, 4
1, 3, 5, 6, 5, 4	1, 3, 5, 6, 6, 4	1, 3, 5, 6, 7, 4	1, 3, 5, 7, 6, 4
1, 3, 5, 7, 7, 4	1, 3, 5, 7, 8, 4	1, 3, 5, 7, 9, 4	1, 3, 6, 6, 5, 4
1, 3, 6, 6, 6, 4	1, 3, 6, 7, 6, 4	1, 3, 6, 7, 7, 4	1, 3, 6, 7, 8, 4
1, 3, 6, 8, 6, 4	1, 3, 6, 8, 7, 4	1, 3, 6, 8, 8, 4	1, 3, 6, 8, 9, 4
1, 3, 6, 8, 10, 4	1, 3, 6, 9, 6, 4	1, 3, 6, 9, 7, 4	1, 3, 6, 9, 8, 4
1, 3, 6, 9, 9, 4	1, 3, 6, 9, 10, 4	1, 3, 6, 9, 11, 4	1, 3, 6, 9, 12, 4
1, 3, 6, 10, 7, 4	1, 3, 6, 10, 8, 4	1, 3, 6, 10, 9, 4	1, 3, 6, 10, 10, 4
1, 3, 6, 10, 11, 4	1, 3, 6, 10, 12, 4		

Table 5.5

1, 3, 4, 5, 5, 5	1, 3, 4, 5, 6, 5	1, 3, 5, 5, 5, 5	1, 3, 5, 6, 6, 5
1, 3, 5, 6, 7, 5	1, 3, 5, 7, 6, 5	1, 3, 5, 7, 7, 5	1, 3, 5, 7, 8, 5
1, 3, 5, 7, 9, 5	1, 3, 6, 6, 6, 5	1, 3, 6, 7, 6, 5	1, 3, 6, 7, 7, 5
1, 3, 6, 7, 8, 5	1, 3, 6, 8, 7, 5	1, 3, 6, 8, 8, 5	1, 3, 6, 8, 9, 5
1, 3, 6, 8, 10, 5	1, 3, 6, 9, 7, 5	1, 3, 6, 9, 8, 5	1, 3, 6, 9, 9, 5
1, 3, 6, 9, 10, 5	1, 3, 6, 9, 11, 5	1, 3, 6, 9, 12, 5	1, 3, 6, 10, 7, 5
1, 3, 6, 10, 8, 5	1, 3, 6, 10, 9, 5	1, 3, 6, 10, 10, 5	1, 3, 6, 10, 11, 5
1, 3, 6, 10, 12, 5	1, 3, 6, 10, 13, 5	1, 3, 6, 10, 14, 5	1, 3, 6, 10, 15, 5

Table 5.6

1, 3, 4, 5, 6, 6	1, 3, 5, 6, 6, 6	1, 3, 5, 6, 7, 6	1, 3, 5, 7, 7, 6
1, 3, 5, 7, 8, 6	1, 3, 5, 7, 9, 6	1, 3, 6, 6, 6, 6	1, 3, 6, 7, 7, 6
1, 3, 6, 7, 8, 6	1, 3, 6, 8, 7, 6	1, 3, 6, 8, 8, 6	1, 3, 6, 8, 9, 6
1, 3, 6, 8, 10, 6	1, 3, 6, 9, 8, 6	1, 3, 6, 9, 9, 6	1, 3, 6, 9, 10, 6
1, 3, 6, 9, 11, 6	1, 3, 6, 9, 12, 6	1, 3, 6, 10, 8, 6	1, 3, 6, 10, 9, 6
1, 3, 6, 10, 10, 6	1, 3, 6, 10, 11, 6	1, 3, 6, 10, 12, 6	1, 3, 6, 10, 13, 6
1, 3, 6, 10, 14, 6	1, 3, 6, 10, 15, 6		

Table 5.7

1, 3, 4, 5, 6, 7	1, 3, 5, 6, 7, 7	1, 3, 5, 7, 7, 7	1, 3, 5, 7, 8, 7
1, 3, 5, 7, 9, 7	1, 3, 6, 7, 7, 7	1, 3, 6, 7, 8, 7	1, 3, 6, 8, 8, 7
1, 3, 6, 8, 9, 7	1, 3, 6, 8, 10, 7	1, 3, 6, 9, 8, 7	1, 3, 6, 9, 9, 7
1, 3, 6, 9, 10, 7	1, 3, 6, 9, 11, 7	1, 3, 6, 9, 12, 7	1, 3, 6, 10, 9, 7
1, 3, 6, 10, 10, 7	1, 3, 6, 10, 11, 7	1, 3, 6, 10, 12, 7	1, 3, 6, 10, 13, 7
1, 3, 6, 10, 14, 7	1, 3, 6, 10, 15, 7		

Table 5.8

1, 3, 5, 6, 7, 8	1, 3, 5, 7, 8, 8	1, 3, 5, 7, 9, 8	1, 3, 6, 7, 8, 8
1, 3, 6, 8, 8, 8	1, 3, 6, 8, 9, 8	1, 3, 6, 8, 10, 8	1, 3, 6, 9, 9, 8
1, 3, 6, 9, 10, 8	1, 3, 6, 9, 11, 8	1, 3, 6, 9, 12, 8	1, 3, 6, 10, 9, 8
1, 3, 6, 10, 10, 8	1, 3, 6, 10, 11, 8	1, 3, 6, 10, 12, 8	1, 3, 6, 10, 13, 8
1, 3, 6, 10, 14, 8	1, 3, 6, 10, 15, 8		

Table 5.9

1, 3, 5, 7, 8, 9	1, 3, 5, 7, 9, 9	1, 3, 6, 7, 8, 9	1, 3, 6, 8, 9, 9
1, 3, 6, 8, 10, 9	1, 3, 6, 9, 9, 9	1, 3, 6, 9, 10, 9	1, 3, 6, 9, 11, 9
1, 3, 6, 9, 12, 9	1, 3, 6, 10, 10, 9	1, 3, 6, 10, 11, 9	1, 3, 6, 10, 12, 9
1, 3, 6, 10, 13, 9	1, 3, 6, 10, 14, 9	1, 3, 6, 10, 15, 9	

Table 5.10

1, 3, 5, 7, 9, 10	1, 3, 6, 8, 9, 10	1, 3, 6, 8, 10, 10	1, 3, 6, 9, 10, 10
1, 3, 6, 9, 11, 10	1, 3, 6, 9, 12, 10	1, 3, 6, 10, 10, 10	1, 3, 6, 10, 11, 10
1, 3, 6, 10, 12, 10	1, 3, 6, 10, 13, 10	1, 3, 6, 10, 14, 10	1, 3, 6, 10, 15, 10

Table 5.11

1, 3, 5, 7, 9, 11	1, 3, 6, 8, 10, 11	1, 3, 6, 9, 10, 11	1, 3, 6, 9, 11, 11
1, 3, 6, 9, 12, 11	1, 3, 6, 10, 11, 11	1, 3, 6, 10, 12, 11	1, 3, 6, 10, 13, 11
1, 3, 6, 10, 14, 11	1, 3, 6, 10, 15, 11		

Table 5.12

1, 3, 6, 8, 10, 12	1, 3, 6, 9, 11, 12	1, 3, 6, 9, 12, 12	1, 3, 6, 10, 11, 12
1, 3, 6, 10, 12, 12	1, 3, 6, 10, 13, 12	1, 3, 6, 10, 14, 12	1, 3, 6, 10, 15, 12

Table 5.13

1, 3, 6, 9, 11, 13	1, 3, 6, 9, 12, 13	1, 3, 6, 10, 13, 13
1, 3, 6, 10, 14, 13	1, 3, 6, 10, 15, 13	

Table 5.14

1, 3, 6, 9, 12, 14	1, 3, 6, 10, 12, 14	1, 3, 6, 10, 13, 14
1, 3, 6, 10, 14, 14	1, 3, 6, 10, 15, 14	

Table 5.15

1, 3, 6, 9, 12, 15	1, 3, 6, 10, 13, 15	1, 3, 6, 10, 14, 15	1, 3, 6, 10, 15, 15

Table 5.16

1, 3, 6, 10, 13, 16	1, 3, 6, 10, 14, 16	1, 3, 6, 10, 15, 16

Table 5.17

1, 3, 6, 10, 14, 17
1, 3, 6, 10, 15, 17

Table 5.18

1, 3, 6, 10, 14, 18
1, 3, 6, 10, 15, 18

Table 5.19

1, 3, 6, 10, 15, 19

Table 5.20

1, 3, 6, 10, 15, 20

Table 5.21

1, 3, 6, 10, 15, 21

Table 6.2

1, 3, 3, 3, 3, 3, 2	1, 3, 4, 4, 4, 3, 2	1, 3, 4, 4, 4, 4, 2
1, 3, 4, 5, 4, 3, 2	1, 3, 4, 5, 5, 4, 2	1, 3, 4, 5, 5, 5, 2
1, 3, 4, 5, 6, 5, 2	1, 3, 5, 5, 5, 4, 2	1, 3, 5, 5, 5, 5, 2
1, 3, 5, 6, 5, 4, 2	1, 3, 5, 6, 6, 4, 2	1, 3, 5, 6, 6, 5, 2
1, 3, 5, 6, 6, 6, 2	1, 3, 5, 6, 7, 5, 2	1, 3, 5, 6, 7, 6, 2
1, 3, 5, 7, 6, 4, 2	1, 3, 5, 7, 6, 5, 2	1, 3, 5, 7, 7, 5, 2
1, 3, 5, 7, 7, 6, 2	1, 3, 5, 7, 8, 5, 2	1, 3, 5, 7, 8, 6, 2
1, 3, 5, 7, 9, 6, 2	1, 3, 6, 6, 6, 4, 2	1, 3, 6, 6, 6, 5, 2
1, 3, 6, 6, 6, 6, 2	1, 3, 6, 7, 6, 4, 2	1, 3, 6, 7, 6, 5, 2
1, 3, 6, 7, 7, 4, 2	1, 3, 6, 7, 7, 5, 2	1, 3, 6, 7, 7, 6, 2
1, 3, 6, 7, 8, 5, 2	1, 3, 6, 7, 8, 6, 2	1, 3, 6, 8, 6, 4, 2
1, 3, 6, 8, 7, 4, 2	1, 3, 6, 8, 7, 5, 2	1, 3, 6, 8, 7, 6, 2
1, 3, 6, 8, 8, 5, 2	1, 3, 6, 8, 8, 6, 2	1, 3, 6, 8, 9, 5, 2
1, 3, 6, 8, 9, 6, 2	1, 3, 6, 8, 10, 6, 2	1, 3, 6, 9, 7, 4, 2
1, 3, 6, 9, 7, 5, 2	1, 3, 6, 9, 8, 5, 2	1, 3, 6, 9, 8, 6, 2
1, 3, 6, 9, 9, 5, 2	1, 3, 6, 9, 9, 6, 2	1, 3, 6, 9, 10, 6, 2
1, 3, 6, 9, 11, 6, 2	1, 3, 6, 9, 12, 6, 2	1, 3, 6, 10, 7, 4, 2
1, 3, 6, 10, 8, 5, 2	1, 3, 6, 10, 8, 6, 2	1, 3, 6, 10, 9, 5, 2
1, 3, 6, 10, 9, 6, 2	1, 3, 6, 10, 10, 6, 2	1, 3, 6, 10, 11, 6, 2
1, 3, 6, 10, 12, 6, 2		

Bibliography

[1] E. Ballico and A.V. Geramita, *The minimal free resolution of the ideal of s general points in P^3*, Proceedings of the 1984 Vancouver conference in algebraic geometry, 1–10, CMS Conf. Proc., 6, Amer. Math. Soc., Providence, RI, 1986.

[2] D. Bayer and M. Stillman, Macaulay: A system for computation in algebraic geometry and commutative algebra. Source and object code available for Unix and Macintosh computers. Contact the authors, or download from ftp://math.harvard.edu via anonymous ftp.

[3] D. Bernstein and A. Iarrobino, *A nonunimodal graded Gorenstein Artin Algebra in Codimension 5*. Comm. Algebra **20**:2323–2336 (1992).

[4] A. M. Bigatti, *Upper Bounds for the Betti Numbers of a Given Hilbert Function*. Comm. Algebra **21**(7):2317–2334 (1993).

[5] A. Bigatti and A. V. Geramita, *Lex-Segments, Level Algebras and Minimal Hilbert Functions*. Comm. Algebra **31**:1427–1451 (2003).

[6] A. Bigatti, A.V. Geramita and J. Migliore, *Geometric Consequences of Extremal Behavior in a Theorem of Macaulay*. Trans. Amer. Math. Soc. **346**:203–235 (1994).

[7] M. Boij, *Artin Level Algebras*. Doctoral Dissertation. Royal Institute of Technology, Stockholm, Sweden, (1994).

[8] M. Boij, *Gorenstein Artin Algebras and Points in Projective Space*. Bull. Lond. Math. Soc. **31**:11–16 (1999).

[9] M. Boij, *Components of the space parametrizing graded Gorenstein Artinian algebras with a given Hilbert function*. Pacific J. Math. **187**:1–11 (1999).

[10] M. Boij, *Graded Gorenstein Artin algebras whose Hilbert functions have a large number of valleys*. Comm. Algebra **23**(1):97–103 (1995).

[11] M. Boij and D. Laksov, *Nonunimodality of graded Gorenstein Artin algebras*. Proc. Amer. Math. Soc. **120**:1083–1092 (1994).

[12] A. Capani, G. Niesi, and L. Robbiano, *CoCoA, a system for doing computations in commutative algebra*. Available via anonymous ftp, cocoa.dima.unige.it.

[13] J.V. Chipalkatti and A.V. Geramita, *On parameter spaces for Artin level algebras*. Michigan Math. J. **51**(1):187–207 (2003).

[14] Y. Cho and A. Iarrobino, *Hilbert Functions of Level Algebras*. J. Algebra **241**:745–758 (2001).

[15] E. Davis, A. V. Geramita and F. Orecchia, *Gorenstein algebras and the Cayley-Bacharach theorem*. Proc. Amer. Math. Soc. **93**:593–597 (1985).

[16] E. DeNegri and G. Valla, *The h-vector of a Gorenstein codimension three domain*. Nagoya Math. J. **138**:113-140 (1995).

[17] S. J. Diesel. *Some irreducibility and dimension theorems for families of height 3 Gorenstein algebras*. Pacific J. Math. **172**:365–397 (1996).

[18] D. Eisenbud and S. Popescu, *Gale Duality and Free Resolutions of Ideals of Points*. Invent. math. **136**:419–449 (1999).

[19] S. Eliahou and M. Kervaire, *Minimal resolutions of some monomial ideals*. J. Algebra **129**:1–25 (1990).

[20] F. Fröberg and D. Laksov, *Compressed Algebras, Complete Intersections (Acireale, 1983)*. Lecture Notes in Math., Springer-Verlag, **1092**:121–151 (1984).

[21] A. V. Geramita, *Waring's Problem for Forms: inverse systems of fat points, secant varieties and Gorenstein algebras*. Queen's Papers in Pure and Applied Math. The Curves Seminar, Vol. X. 105 (1996).

[22] A.V. Geramita, D. Gregory and L. Roberts, *Monomial Ideals and Points in Projective Space*. J. Pure Appl. Algebra **40**:33-62 (1986).

[23] A. V. Geramita, T. Harima, and Y. S. Shin, *Some special configurations of points in $\mathbb{P}^n$*. J. Algebra **268**:484–518 (2003)
[24] A. V. Geramita and A. Lorenzini, *Cancelable Hilbert functions and Level Algebras*. Comm. Algebra (to appear).
[25] A. V. Geramita, P. Maroscia, and L. Roberts, *The Hilbert function of a reduced K-algebra*. J. London Math. Soc. **28**:443–452, (1983).
[26] A. V. Geramita and J. C. Migliore, *Hyperplane sections of a smooth curve in $\mathbb{P}^3$*. Comm. Algebra **17**(12):3129–3164 (1989).
[27] A. V. Geramita and J. C. Migliore, *A Generalized Liaison Addition*. J. Algebra **163**:139–164 (1994).
[28] A. V. Geramita and J. C. Migliore, *Reduced Gorenstein Codimension 3 subschemes of projective space*. Proc. Amer. Math. Soc. **125**:943–950 (1997).
[29] A. V. Geramita, M. Pucci, and Y. S. Shin, *Smooth Points of $\mathcal{G}or(T)$*. J. Pure Appl. Algebra. **122**:209–241 (1997).
[30] D. R. Grayson and M. E. Stillman, *Macaulay 2: A system for computation in algebraic geometry and commutative algebra*. Contact the authors, or download from www.math.uiuc.edu/Macaulay2.
[31] T. Harima, *Some Examples of unimodal Gorenstein sequences*. J. Pure Appl. Algebra. **103**:313–324 (1995).
[32] T. Harima, *Characterization of Hilbert functions of Gorenstein Artin algebras with the weak Stanley property*. Proc. Amer. Math. Soc. **123**:3631–3638 (1995).
[33] T. Harima, *A Note on Artinian Gorenstein Algebras of Codimension Three*. J. Pure Appl. Algebra. **135**:45–56 (1999).
[34] J. Harris, *The genus of space curves*. Math. Annalen **249**:191–204 (1980).
[35] T. Harima, J. C. Migliore, U. Nagel, and J. Watanabe, *The weak and strong Lefschetz properties for Artinian K-algebras*. J. Algebra, **262**:99-126 (2003).
[36] R. Hartshorne, *Some Examples of Gorenstein Liaison in codimension 3*. Collect. Math. **53**(1):21–48 (2002).
[37] J. Herzog, N. V. Trung, and G. Valla, *On the Hyperplane Sections of Reduced, Irreducible Varieties of Low Codimension*. J. Math. Kyoto Univ. **34**:47–72 (1994).
[38] T. Hibi, *Flawless O-sequences and Hilbert functions of Cohen-Macaulay Integral Domains*. J. Pure Appl. Algebra **60**:245–251 (1989).
[39] A. Hirschowitz, and C. Simpson, *La Résolution minimale de l'idéal d'un arrangement général d'un grand nombre de points dans $\mathbb{P}^n$*. Invent. math. **126**:467–503 (1996).
[40] M. Hochster and D. Laksov, *The Linear Syzygies of Generic Forms*. Comm. Algebra **15**:227–239 (1987).
[41] H. A. Hulett, *Maximum Betti Numbers of Homogeneous Ideals with a Given Hilbert Function*. Comm. Algebra. **21**(7):2335–2350 (1993).
[42] C. Huneke, *Numerical Invariants of Liaison classes*. Invent. math. **75**:301–325 (1984).
[43] A. Iarrobino, *Compressed Algebras: Artin Algebras having given Socle degrees and Maximal Length*. Trans. Amer. Math. Soc. **285**:337–378 (1984).
[44] A. Iarrobino, *Ancestor ideals of vector spaces of forms, and level algebras*. J. Algebra **272**:530–580 (2004).
[45] A. Iarrobino, *Hilbert functions of Gorenstein algebras associated to a pencil of forms*, preprint.
[46] A. Iarrobino and V. Kanev, *Power Sums, Gorenstein Algebras and Determinantal Loci*. Lecture Notes in Math., Springer-Verlag, **1721** (1999).
[47] H. Ikeda, *Results on Dilworth and Rees numbes of Artinian local rings*. Japan. J. Math. **22**:147–158 (1996).
[48] J. Kleppe, J. C. Migliore, R. M. Miro-Roig, U. Nagel, and C. Peterson, *Gorenstein Liaison, Complete Intersection Liaison, Invariants and Unobstructedness*. Mem. Amer. Math. Soc. **154** (2000).
[49] F. Lauze, *Rang maximal pour $T_{\mathbb{P}^n}$*. Manuscripta Math. **92**:525–543 (1997).
[50] A. Lorenzini, *The Minimal Resolution Conjecture*. J. Algebra **156**:5–35 (1993).
[51] F. H. S. Macaulay, *Some properties of enumeration in the theory of modular systems*. Proc. Lond. Math. Soc. **26**:531–555 (1927).
[52] R. Maggioni and A. Ragusa, *The Hilbert function of generic plane sections of curves in $\mathbb{P}^3$*. Invent. Math. **91**:253–258 (1988).

[53] J. C. Migliore, *Introduction to Liaison Theory and Deficiency Modules*. Birkhäuser, Progress in Math. **165** (1988), 224 pp.

[54] J.C. Migliore and R. Miró-Roig, *Ideals of general forms and the ubiquity of the Weak Lefschetz Property*. J. Pure Appl. Algebra **182**:79–107 (2003).

[55] J. C. Migliore and U. Nagel, *Reduced arithmetically Gorenstein schemes and Simplicial polytopes with maximal Betti numbers*. Adv. Math. **180**:1–63 (2003).

[56] J. C. Migliore and U. Nagel, *Liaison and Related Topics: Notes from the Torino Workshop/School*. Rend. Sem. Mat.-Torino **59(2)**:59–126 (2001).

[57] J.C. Migliore, W. Nagel, and C. Peterson, *Buchsbaum-Rim sheaves and their multiple sections*. J. Algebra **219**:378–420 (1999).

[58] J.C. Migliore and C. Peterson, *A construction of codimension three arithmetically Gorenstein subschemes of projective space*. Trans. Amer. Math. Soc. **349**:3803–3821 (1997).

[59] Mustaţă, Mircea, *Graded Betti numbers of general finite subsets of points on projective varieties*. Pragmatic 1997 (Catania). Le Matematiche (Catania) **53**: (1998)(suppl. 1999).

[60] K. Pardue, *Deformation Classes of Graded Modules and Maximal Betti Numbers*. Illinois J. Math. **40**:564–585 (1996).

[61] I. Peeva, *Consecutive cancelations in Betti numbers*. Proc. Amer. Math. Soc. **132**:3503–3507 (2004).

[62] A. Ragusa and G. Zappala, *Partial Intersections and Graded Betti Numbers*. Beiträge Algebra Geom. **44(1)**:285–302 (2003).

[63] R. Stanley, *Cohen-Macaulay Complexes*. In: *Higher combinatorics (Proc. NATO Advanced Study Inst., Berlin, 1976)*, pp. 51–62. NATO Adv. Study Inst. Ser., Ser. C: Math. and Phys. Sci., 31. Reidel, Dordrecht, 1977.

[64] R. Stanley, *Hilbert Functions of Graded Algebras*. Adv. Math., **28**:57–83 (1978).

[65] R. Stanley, *On the Hilbert function of a graded Cohen-Macaulay domain*. J. Pure Appl. Algebra **73**:307–314 (1991).

[66] R. Stanley, *Cohen-Macaulay rings and constructible polytopes*. Bull. Amer. Math. Soc. **81**:133–135 (1975).

[67] R. Stanley, *The number of faces of a simplicial convex polytope*, Advances in Math. **35** (1980), 236–238,

[68] R. Stanley, *The number of faces of simplicial polytopes and spheres*. In: *Discrete geometry and convexity (New York, 1982)*, 212–223, Ann. New York Acad. Sci., 440, New York Acad. Sci., New York, 1985

[69] C. Walter, *The minimal free resolution of the homogeneous ideal of s general points in $\mathbb{P}^4$*. Math. Zeit. **219(2)**:231–234 (1995).

[70] J. Watanabe, *A note on Gorenstein rings of embedding codimension 3*. Nagoya Math. J. **50**:227–282 (1973).

[71] J. Watanabe, *The Dilworth number of Artinian rings and finite posets with rank function*. Avd. Stud. Pure Math. **11**:303–312 (1973).

[72] J. Watanabe, *A Note on Complete intersections of Height Three*. Proc. Amer. Math. Soc. **126**:3161–3168 (1973).

Editorial Information

To be published in the *Memoirs*, a paper must be correct, new, nontrivial, and significant. Further, it must be well written and of interest to a substantial number of mathematicians. Piecemeal results, such as an inconclusive step toward an unproved major theorem or a minor variation on a known result, are in general not acceptable for publication.

Papers appearing in *Memoirs* are generally at least 80 and not more than 200 published pages in length. Papers less than 80 or more than 200 published pages require the approval of the Managing Editor of the Transactions/Memoirs Editorial Board.

As of November 30, 2006, the backlog for this journal was approximately 12 volumes. This estimate is the result of dividing the number of manuscripts for this journal in the Providence office that have not yet gone to the printer on the above date by the average number of monographs per volume over the previous twelve months, reduced by the number of volumes published in four months (the time necessary for preparing a volume for the printer). (There are 6 volumes per year, each usually containing at least 4 numbers.)

A Consent to Publish and Copyright Agreement is required before a paper will be published in the *Memoirs*. After a paper is accepted for publication, the Providence office will send a Consent to Publish and Copyright Agreement to all authors of the paper. By submitting a paper to the *Memoirs*, authors certify that the results have not been submitted to nor are they under consideration for publication by another journal, conference proceedings, or similar publication.

Information for Authors

Memoirs are printed from camera copy fully prepared by the author. This means that the finished book will look exactly like the copy submitted.

Initial submission. The AMS uses Centralized Manuscript Processing for initial submissions. Authors should submit a PDF file using the Initial Manuscript Submission form found at www.ams.org/cgi-bin/peertrack/submission.pl, or send one copy of the manuscript to the following address: Centralized Manuscript Processing, MEMOIRS OF THE AMS, 201 Charles Street, Providence, RI 02904-2294 USA. If a paper copy is being forwarded to the AMS, indicate that it is for it Memoirs and include the name of the corresponding author, contact information such as email address or mailing address, and the name of an appropriate Editor to review the paper (see the list of Editors below).

The paper must contain a *descriptive title* and an *abstract* that summarizes the article in language suitable for workers in the general field (algebra, analysis, etc.). The *descriptive title* should be short, but informative; useless or vague phrases such as "some remarks about" or "concerning" should be avoided. The *abstract* should be at least one complete sentence, and at most 300 words. Included with the footnotes to the paper should be the 2000 *Mathematics Subject Classification* representing the primary and secondary subjects of the article. The classifications are accessible from www.ams.org/msc/. The list of classifications is also available in print starting with the 1999 annual index of *Mathematical Reviews*. The Mathematics Subject Classification footnote may be followed by a list of *key words and phrases* describing the subject matter of the article and taken from it. Journal abbreviations used in bibliographies are listed in the latest *Mathematical Reviews* annual index. The series abbreviations are also accessible from www.ams.org/publications/. To help in preparing and verifying references, the AMS offers MR Lookup, a Reference Tool for Linking, at www.ams.org/mrlookup/.

Electronically prepared manuscripts. The AMS encourages electronically prepared manuscripts, with a strong preference for $\mathcal{A}_{\mathcal{M}}\mathcal{S}$-LaTeX. To this end, the Society has prepared $\mathcal{A}_{\mathcal{M}}\mathcal{S}$-LaTeX author packages for each AMS publication. Author packages include instructions for preparing electronic manuscripts, samples, and a style file that generates the particular design specifications of that publication series. Though $\mathcal{A}_{\mathcal{M}}\mathcal{S}$-LaTeX is the highly preferred format of TeX, author packages are also available in $\mathcal{A}_{\mathcal{M}}\mathcal{S}$-TeX.

Authors may retrieve an author package from the AMS website starting from www.ams.org/tex/ or via FTP to ftp.ams.org (login as anonymous, enter username as password, and type cd pub/author-info). The *AMS Author Handbook* and the *Instruction Manual* are available in PDF format following the author packages link from www.ams.org/tex/. The author package can also be obtained free of charge by sending email to tech-support@ams.org (Internet) or from the Publication Division, American Mathematical Society, 201 Charles St., Providence, RI 02904-2294, USA. When requesting an author package, please specify $\mathcal{A}_{\mathcal{M}}\mathcal{S}$-LaTeX or $\mathcal{A}_{\mathcal{M}}\mathcal{S}$-TeX and the publication in which your paper will appear. Please be sure to include your complete mailing address.

After acceptance. The final version of the electronic file should be sent to the Providence office (this includes any TeX source file, any graphics files, and the DVI or PostScript file) immediately after the paper has been accepted for publication.

Before sending the source file, be sure you have proofread your paper carefully. The files you send must be the EXACT files used to generate the proof copy that was accepted for publication. For all publications, authors are required to send a printed copy of their paper, which exactly matches the copy approved for publication, along with any graphics that will appear in the paper.

Accepted electronically prepared files can be submitted via the web at www.ams.org/submit-book-journal/, sent via FTP, or sent on CD-Rom or diskette to the Electronic Prepress Department, American Mathematical Society, 201 Charles Street, Providence, RI 02904-2294 USA. TeX source files, DVI files, and PostScript files can be transferred over the Internet by FTP to the Internet node ftp.ams.org (130.44.1.100). When sending a manuscript electronically via CD-Rom or diskette, please be sure to include a message identifying the paper as a Memoir.

Electronically prepared manuscripts can also be sent via email to pub-submit@ams.org (Internet). In order to send files via email, they must be encoded properly. (DVI files are binary and PostScript files tend to be very large.)

Electronic graphics. Comprehensive instructions on preparing graphics are available at www.ams.org/jourhtml/. A few of the major requirements are given here.

Submit files for graphics as EPS (Encapsulated PostScript) files. This includes graphics originated via a graphics application as well as scanned photographs or other computer-generated images. If this is not possible, TIFF files are acceptable as long as they can be opened in Adobe Photoshop or Illustrator. No matter what method was used to produce the graphic, it is necessary to provide a paper copy to the AMS.

Authors using graphics packages for the creation of electronic art should also avoid the use of any lines thinner than 0.5 points in width. Many graphics packages allow the user to specify a "hairline" for a very thin line. Hairlines often look acceptable when proofed on a typical laser printer. However, when produced on a high-resolution laser imagesetter, hairlines become nearly invisible and will be lost entirely in the final printing process.

Screens should be set to values between 15% and 85%. Screens which fall outside of this range are too light or too dark to print correctly. Variations of screens within a graphic should be no less than 10%.

Inquiries. Any inquiries concerning a paper that has been accepted for publication should be sent to memo-query@ams.org or directly to the Electronic Prepress Department, American Mathematical Society, 201 Charles St., Providence, RI 02904-2294 USA.

Editors

This journal is designed particularly for long research papers, normally at least 80 pages in length, and groups of cognate papers in pure and applied mathematics. Papers intended for publication in the *Memoirs* should be addressed to one of the following editors. The AMS uses Centralized Manuscript Processing for initial submissions to AMS journals. Authors should follow instructions listed on the Initial Submission page found at www.ams.org/memo/memosubmit.html.

Algebra to ALEXANDER KLESHCHEV, Department of Mathematics, University of Oregon, Eugene, OR 97403-1222; email: ams@noether.uoregon.edu

Algebra and its application to MINA TEICHER, Emmy Noether Research Institute for Mathematics, Bar-Ilan University, Ramat-Gan 52900, Israel; email: teicher@macs.biu.ac.il

Algebraic geometry to DAN ABRAMOVICH, Department of Mathematics, Brown University, Box 1917, Providence, RI 02912; email: amsedit@math.brown.edu

Algebraic number theory to V. KUMAR MURTY, Department of Mathematics, University of Toronto, 100 St. George Street, Toronto, ON M5S 1A1, Canada; email: murty@math.toronto.edu

Algebraic topology to ALEJANDRO ADEM, Department of Mathematics, University of British Columbia, Room 121, 1984 Mathematics Road, Vancouver, British Columbia, Canada V6T 1Z2; email: adem@math.ubc.ca

Combinatorics to JOHN R. STEMBRIDGE, Department of Mathematics, University of Michigan, Ann Arbor, Michigan 48109-1109; email: FRS@umich.edu

Complex analysis and harmonic analysis to ALEXANDER NAGEL, Department of Mathematics, University of Wisconsin, 480 Lincoln Drive, Madison, WI 53706-1313; email: nagel@math.wisc.edu

Differential geometry and global analysis to LISA C. JEFFREY, Department of Mathematics, University of Toronto, 100 St. George St., Toronto, ON Canada M5S 3G3; email: jeffrey@math.toronto.edu

Dynamical systems and ergodic theory to AMIE WILKINSON, Department of Mathematics, Northwestern University, 2033 Sheridan Road, Evanston, IL 60208-2730; email: transactions@math.northwestern.edu

Functional analysis and operator algebras to DIMITRI SHLYAKHTENKO, Department of Mathematics, University of California, Los Angeles, CA 90095; email: shlyakht@math.ucla.edu

Geometric analysis to WILLIAM P. MINICOZZI II, Department of Mathematics, Johns Hopkins University, 3400 N. Charles St., Baltimore, MD 21218; email: trans@math.jhu.edu

Geometric analysis to MLADEN BESTVINA, Department of Mathematics, University of Utah, 155 South 1400 East, JWB 233, Salt Lake City, Utah 84112-0090; email: bestvina@math.utah.edu

Harmonic analysis, representation theory, and Lie theory to ROBERT J. STANTON, Department of Mathematics, The Ohio State University, 231 West 18th Avenue, Columbus, OH 43210-1174; email: stanton@math.ohio-state.edu

Logic to STEFFEN LEMPP, Department of Mathematics, University of Wisconsin, 480 Lincoln Drive, Madison, Wisconsin 53706-1388; email: lempp@math.wisc.edu

Partial differential equations to GUSTAVO PONCE, Department of Mathematics, South Hall, Room 6607, University of California, Santa Barbara, CA 93106; email: ponce@math.ucsb.edu

Partial differential equations and dynamical systems to PETER POLACIK, School of Mathematics, University of Minnesota, Minneapolis, MN 55455; email: polacik@math.umn.edu

Probability and statistics to KRZYSZTOF BURDZY, Department of Mathematics, University of Washington, Box 354350, Seattle, Washington 98195-4350; email: burdzy@math.washington.edu

Real analysis and partial differential equations to DANIEL TATARU, Department of Mathematics, University of California, Berkeley, Berkeley, CA 94720; email: tataru@math.berkeley.edu

All other communications to the editors should be addressed to the Managing Editor, ROBERT GURALNICK, Department of Mathematics, University of Southern California, Los Angeles, CA 90089-1113; email: guralnic@math.usc.edu.

Titles in This Series

875 **C. Krattenthaler and T. Rivoal,** Hypergéométrie et fonction zêta de Riemann, 2007

874 **Sonia Natale,** Semisolvability of semisimple Hopf algebras of low dimension, 2007

873 **A. J. Duncan,** Exponential genus problems in one-relator products of groups, 2007

872 **Anthony V. Geramita, Tadahito Harima, Juan C. Migliore, and Yong Su Shin,** The Hilbert function of a level algebra, 2007

871 **Pascal Auscher,** On necessary and sufficient conditions for L^p-estimates of Riesz transforms associated to elliptic operators on $\mathbb{R}^n$ and related estimates, 2007

870 **Takuro Mochizuki,** Asymptotic behaviour of tame harmonic bundles and an application to pure twistor D-modules, Part 2, 2007

869 **Takuro Mochizuki,** Asymptotic behaviour of tame harmonic bundles and an application to pure twistor D-modules, Part 1, 2007

868 **Gelu Popescu,** Entropy and multivariable interpolation, 2006

867 **Vilmos Totik,** Metric properties of harmonic measures, 2006

866 **William Craig,** Semigroups underlying first-order logic, 2006

865 **Nathanial P. Brown,** Invariant means and finite representation theory of $C*$-algebras, 2006

864 **John M. Lee,** Fredholm operators and Einstein metrics on conformally compact manifolds, 2006

863 **M. Lübke and A. Teleman,** The Universal Kobayashi-Hitchin correspondence on Hermitian manifolds, 2006

862 **Alberto Canonaco,** The Beilinson complex and canonical rings of irregular surfaces, 2006

861 **Leon A. Takhtajan and Lee-Peng Teo,** Weil-Petersson metric on the universal Teichmüller space, 2006

860 **Thomas M. Fiore,** Pseudo limits, biadjoints and pseudo algebras: Categorical foundations of conformal field theory, 2006

859 **N. Arcozzi, R. Rochberg, and E. Sawyer,** Carleson measures and interpolating sequences for Besov spaces on complex balls, 2006

858 **Enrico Valdinoci, Berardino Sciunzi, and Vasile Ovidiu Savin,** Flat level set regularity of p-Laplace phase transitions, 2006

857 **Donatella Danielli, Nocola Garofalo, and Duy-Minh Nhieu,** Non-doubling Ahlfors measures, perimeter measures, and the characterization of the trace spaces of Sobolev functions in Carnot-Carathéodory spaces, 2006

856 **Vladimir Bolotnikov and Harry Dym,** On boundary interpolation for matrix valued Schur functions, 2006

855 **Yevgenia Kashina, Yorck Sommerhäuser, and Yongchang Zhu,** On higher Frobenius-Schur indicators, 2006

854 **Noam Greenberg,** The role of true finiteness in the admissible recursively enumerable degrees, 2006

853 **Joachim Krieger,** Stability of spherically symmetric wave maps, 2006

852 **Viorel Barbu, Irena Lasiecka, and Roberto Triggiani,** Tangential boundary stabilization of Navier-Stokes equations, 2006

851 **Jie Wu,** On maps from loop suspensions to loop spaces and the shuffle relations on the Cohen groups, 2006

850 **Siegfried Echterhoff, S. Kaliszewski, John Quigg, and Iain Raeburn,** A categorical approach to imprimitivity theorems for C^*-dynamical systems, 2006

849 **Katsuhiko Kuribayashi, Mamoru Mimura, and Tetsu Nishimoto,** Twisted tensor products related to the cohomology of the classifying spaces of loop groups, 2006

848 **Bob Oliver,** Equivalences of classifying spaces completed at the prime two, 2006

TITLES IN THIS SERIES

847 Eric T. Sawyer and Richard L. Wheeden, Hölder continuity of weak solutions to subelliptic equations with rough coefficients, 2006

846 Victor Beresnevich, Detta Dickinson, and Sanju Velani, Measure theoretic laws for lim–sup sets, 2006

845 Ehud Friedgut, Vojtech Rödl, Andrzej Ruciński, and Prasad V. Tetali, A Sharp threshold for random graphs with a monochromatic triangle in every edge coloring, 2006

844 Amadeu Delshams, Rafael de la Llave, and Tere M. Seara, A geometric mechanism for diffusion in Hamiltonian systems overcoming the large gap problem: Heuristics and rigorous verification on a model, 2006

843 Denis V. Osin, Relatively hyperbolic groups: Intrinsic geometry, algebraic properties, and algorithmic problems, 2006

842 David P. Blecher and Vrej Zarikian, The calculus of one-sided M-ideals and multipliers in operator spaces, 2006

841 Enrique Artal Bartolo, Pierrette Cassou-Noguès, Ignacio Luengo, and Alejandro Melle Hernández, Quasi-ordinary power series and their zeta functions, 2005

840 Sławomir Kołodziej, The complex Monge-Ampère equation and pluripotential theory, 2005

839 Mihai Ciucu, A random tiling model for two dimensional electrostatics, 2005

838 V. Jurdjevic, Integrable Hamiltonian systems on complex Lie groups, 2005

837 Joseph A. Ball and Victor Vinnikov, Lax-Phillips scattering and conservative linear systems: A Cuntz-algebra multidimensional setting, 2005

836 H. G. Dales and A. T.-M. Lau, The second duals of Beurling algbras, 2005

835 Kiyoshi Igusa, Higher complex torsion and the framing principle, 2005

834 Ken'ichi Ohshika, Kleinian groups which are limits of geometrically finite groups, 2005

833 Greg Hjorth and Alexander S. Kechris, Rigidity theorems for actions of product groups and countable Borel equivalence relations, 2005

832 Lee Klingler and Lawrence S. Levy, Representation type of commutative Noetherian rings III: Global wildness and tameness, 2005

831 K. R. Goodearl and F. Wehrung, The complete dimension theory of partially ordered systems with equivalence and orthogonality, 2005

830 Jason Fulman, Peter M. Neumann, and Cheryl E. Praeger, A generating function approach to the enumeration of matrices in classical groups over finite fields, 2005

829 S. G. Bobkov and B. Zegarlinski, Entropy bounds and isoperimetry, 2005

828 Joel Berman and Paweł M. Idziak, Generative complexity in algebra, 2005

827 Trevor A. Welsh, Fermionic expressions for minimal model Virasoro characters, 2005

826 Guy Métivier and Kevin Zumbrun, Large viscous boundary layers for noncharacteristic nonlinear hyperbolic problems, 2005

825 Yaozhong Hu, Integral transformations and anticipative calculus for fractional Brownian motions, 2005

824 Luen-Chau Li and Serge Parmentier, On dynamical Poisson groupoids I, 2005

823 Claus Mokler, An analogue of a reductive algebraic monoid whose unit group is a Kac-Moody group, 2005

822 Stefano Pigola, Marco Rigoli, and Alberto G. Setti, Maximum principles on Riemannian manifolds and applications, 2005

For a complete list of titles in this series, visit the
AMS Bookstore at **www.ams.org/bookstore/**.